ROUTLEDGE LIBRARY EDITIONS:
HOUSING GENTRIFICATION AND
REGIONAL INEQUALITY

T0201671

Volume 2

REGIONAL POLICY IN BRITAIN

REGIONAL POLICY IN BRITAIN
The North-South Divide

PAUL N. BALCHIN

Routledge
Taylor & Francis Group

LONDON AND NEW YORK

First published in 1990 by Paul Chapman Publishing Ltd

This edition first published in 2022
by Routledge
2 Park Square, Milton Park, Abingdon, Oxon OX14 4RN

and by Routledge
605 Third Avenue, New York, NY 10158

Routledge is an imprint of the Taylor & Francis Group, an informa business

© 1990 Paul N. Balchin

All rights reserved. No part of this book may be reprinted or reproduced or utilised
in any form or by any electronic, mechanical, or other means, now known or
hereafter invented, including photocopying and recording, or in any information
storage or retrieval system, without permission in writing from the publishers.

Trademark notice: Product or corporate names may be trademarks or registered
trademarks, and are used only for identification and explanation without intent to
infringe.

British Library Cataloguing in Publication Data
A catalogue record for this book is available from the British Library

ISBN: 978-1-03-204163-6 (Set)
ISBN: 978-1-00-319078-3 (Set) (ebk)
ISBN: 978-1-03-204134-6 (Volume 2) (hbk)
ISBN: 978-1-03-204160-5 (Volume 2) (pbk)
ISBN: 978-1-00-319071-4 (Volume 2) (ebk)

Publisher's Note
The publisher has gone to great lengths to ensure the quality of this reprint but
points out that some imperfections in the original copies may be apparent.

Disclaimer
The publisher has made every effort to trace copyright holders and would welcome
correspondence from those they have been unable to trace.

REGIONAL POLICY IN BRITAIN

The North-South Divide

PAUL N. BALCHIN

P·C·P

Paul Chapman
Publishing Ltd

Copyright © 1990 Paul Balchin

All rights reserved

Published 1990 by
Paul Chapman Publishing Ltd
144 Liverpool Road
London N1 1LA

No part of this book may be reproduced in any manner whatsoever without the written permission of the publishers, except in the case of brief quotations embodied in critical articles or reviews.

British Library Cataloguing in Publication Data

Balchin, Paul N.
Regional policy in Britain: the north-south divide.
1. Great Britain. Regional economic development. Policies of government, history
I. Title
330.941

ISBN 1 85396 060 8

Typeset by Burns & Smith, Derby
Printed and bound by Butler & Tanner, Frome, Somerset

CONTENTS

Preface   vii

1. INTRODUCTION   1
    Perpetual imbalance   2
    The macroeconomics of Thatcherism   5

2. THE NORTH-SOUTH DIVIDE - REALITY?   7
    Employment - the great divide   7
    Interregional disparities in unemployment   22
    Interregional inequalities in production, incomes and expenditure   29
    Labour-supply disparity   33
    Housing - the ultimate divide?   39
    Unequal health   51
    The prosperity divide   56
    Political polarization   57

3. INTERREGIONAL POLICY   64
    Regional policy: a consensus approach   65
    The phasing-out of regional policy, 1979-87   67
    The enterprise initiative: a policy of confusion   76
    The inner cities   84
    Political consensus and divergence   95

4. THE NORTH-SOUTH DIVIDE - MYTH?   101
    Employment disparity - some contradictions   102
    Unemployment - a re-examination   112
    Inequalities in production and expenditure - a cautionary view   123
    The decentralization of population and employment   129
    Housing: the problems of the south   137
    The health divide: alternative perspectives   144
    The prosperity divide: a reinterpretation   147
    The diversity of electoral support   154

5. POLICY FOR THE 1990s   159
   The north-south divide: policy considerations and options   159
   Policies for the inner cities   168
   Devolution to the regions   171
   Political options   182

   Bibliography and references   185
   Author Index   195
   Subject Index   199

PREFACE

During the depression of the late 1920s and early 1930s, the State began to take an active part in reducing serious interregional imbalances in employment opportunities and economic growth. Whether for reasons of economic efficiency, equity or political expedience, consecutive governments over the subsequent four decades attempted - through a succession of measures - to create a more spatially balanced economy. There was a broad consensus among the political parties that something had to be done to bring about a regional convergence of unemployment rates and growth potential. In adopting 'one-nation' policies, governments - National, Coalition, Labour and Conservative - attempted to eliminate or at least to narrow the 'north-south divide'. Throughout much of the 'Long Boom' (1951-73), Keynesian demand-management policy ensured comparatively low unemployment and a steady rate of economic growth, but these achievements detracted attention from the very serious underlying problems of regional imbalance - although there were imaginative attempts during the Wilson governments of 1964-70, and in the latter part of the Heath administration of 1970-4, to create the right machinery and to provide the means to grapple with the many and complex problems of regional inequality.

With the introduction of monetarist policy in the mid-1970s (in response to both domestic and external pressures), regional assistance was substantially reduced despite rising unemployment and slower economic growth - particularly in the north. With the application of severe deflationary policies by the Thatcher administration in the early 1980s, both the pace of deindustrialization and the rate of unemployment increased -widening still further the north-south divide. In so far as Conservative governments after 1979 broadly perpetuated the fiscal priorities of the previous Labour government, it could be argued that consensus remained intact; but any similarity between policies was largely illusory. Thatcher administrations soon dismantled the regional planning system, rolled back the boundaries of areas eligible for maximum regional development aid and replaced mandatory grants with selective assistance - all with the aim of furthering the evolution of a free-market economy. At an urban level, local authorities were rendered increasingly powerless as rate-support grant and housing subsidies were slashed, rate-capping was imposed and metropolitan counties and the Greater London Council were abolished, while simultaneously an attempt was made to introduce an 'enterprise culture' in the inner cities through the medium of locally unaccountable agencies.

Although much has been written recently on the 'north-south divide', a great deal of policy analysis relates to the mid-1980s or before and rests on the over-generalized assumption that the north is extensively depressed, while the south – by comparison – is a 'land of milk and honey' throughout. Intra-regional disparities are often given scant attention and the urban aspects of regional policy are sometimes neglected. Very rarely is any consideration made of future policy.

This book, therefore, provides a comprehensive and up-to-date review of economic and social data in an attempt to examine whether the north-south divide is a 'reality' or a 'myth'. It seeks to question whether or not public policy has been both effective and consistent over the years (particularly during the period of Thatcher government), and considers changes in regional policy that might emerge in the 1990s. It should appeal to a broad range of university and polytechnic students on degree courses in economics, estate management, geography and town planning, and it is hoped that it could also form the basis for postgraduate study in a relevant subject area while being generally intelligible to someone without a formal background in, say, economics or geography.

Derived in part from *Regional and Urban Economics* (Balchin and Bull, 1986), this text is divided into five chapters. Chapter 1 provides an introductory background to regional imbalance and regional policy. Chapter 2 examines the economic and social data that suggest that there is not only a north-south divide but also that the gap between the two parts of Britain has widened in recent years. Chapter 3 reviews the development of regional policy since the 1930s and considers particularly whether such policy has been both effective and consistent in recent years. Chapter 4 sets out to question some well-known assumptions about regional imbalance and – to an extent – argues that intra-regional disparities might be just as much a serious cause for concern as interregional variation. Chapter 5 concludes by looking at some possible reforms to regional policy.

It should be noted that a number of conventions are used throughout the book. First, the term 'north' should not be confused with the standard region of the North. The former is generic and defined in Chapter 1, and the latter is an official region whose boundaries have been drawn by government. Second, although the term 'north-south divide' applies specifically to Great Britain rather than to the UK as a whole, reference is sometimes made to the UK or Northern Ireland in order to conform with the customary presentation of figures or because of the difficulty in disaggregating published data. Third, throughout this book – and only for the sake of convenience – the term 'region' (rather than any other nomenclature) applies to Scotland, Wales and Northern Ireland, as well as to the English standard regions. Fourth, since government policy is normally judged in the light of official statistics, the unemployment data set out in this book are derived solely from Department of Employment sources (except where otherwise stated). It must be borne in mind, however, that because there were 24 changes (between 1982 and 1989) in the method of counting the unemployed, official statistics are significantly lower that the figures published by the independent Unemployment Unit – which show unemployment according to pre-1982 criteria. In June 1988, for example, whereas unemployment in the UK was 2,375,000 (or 8.4 per cent) according to the official count, it was as much as 3,033,200 (or 10.8 per cent) when calculated by the Unemployment Unit.

I would like to acknowledge a debt I owe to present and former colleagues and to numerous other people (both in the fields of planning and in research) who have stimulated and advised me in the preparation of this book. In particular, I owe debts of gratitude to Brian Gregory of the School of Surveying at Thames Polytechnic, who processed many of the illustrations quickly and efficiently, and to my wife, Maria, not only for her encouragement and patience but also for the many hours she spent at the word-processor.

Paul N. Balchin
London
1989

1

INTRODUCTION

Prior to any examination of the north-south divide, it is necessary to define the terms 'north' and 'south' as used throughout the book. The Town and Country Planning Association (TCPA) (1987) posited that the boundary between the two parts of Britain runs approximately along a line drawn from the Severn to the Wash, but in addition suggested that the west-Midland counties of Hereford & Worcestershire and Warwickshire, and the east-Midland county of North-amptonshire, should be included in the 'south', while the south-western counties of Devon and Cornwall could be regarded as being within the 'north'. To be sure, such a division is plausible if it is determined by fine-grained socio-economic data (for example, at the level of the local labour-market), but when undertaking a study of broad interregional disparities, data relating to the region or county may be more appropriate. There is also the question of 'regional consciousness': it is unlikely, for example, that many people resident in Devon or Cornwall would regard themselves as living in the north.

Partly to avoid the problems of collecting and collating data for split standard regions, recent research into the future of the north-south divide incorporated the whole of the West Midlands and East Midlands into the 'south' (see Cambridge Econometrics and Northern Ireland Economic Research Centre, 1988). Although growth projections suggested that both of the Midlands regions broadly might experience the same rate of growth as the South East, South West and East Anglia to the year 2000, the research ignores current and past similarities between the Midlands and the North West and Yorkshire & Humberside. It would also be very improbable to find many people in, for example, the Potteries of Staffordshire or the Derbyshire coalfield who would claim to be 'southerners'. Nevertheless, with regard to degrees of prosperity, life-style and political preference, the distinction between north and south becomes very blurred. Hackney and Southwark might have more in common with Liverpool or Sunderland than with the outer South East – and as such could be regarded as 'northern' – whereas Harrogate or Macclesfield might appear to be as affluent as

Guildford or Winchester – and could thus be assumed to be 'southern'. In this respect, the TCPA (1987, p. 5) rightly suggested that 'at the end of the day North-South is not so much a geographical concept as a state of mind, brought about by a rapid and catastrophic polarization of the British economy'. None of the above definitions, however, fully offer the possibility of intelligibly examining current north-south disparities, or of considering whether or not the north-south divide is widening or narrowing.

In order to use accessible economic and social data to make clear comparisons between the north and the south, the boundary between these two parts of Britain must, first, avoid dissecting standard regions and, second, divide Britain into two parts that are broadly compatible with people's perception of the terms 'north' and 'south'. To reiterate, it is a nonsense to suggest that whereas Devon is 'northern', Derbyshire is 'southern'. For these reasons it is assumed throughout this book that 'the fault line of [Britain] ... remains, as it has always been, a boundary drawn between the Severn and the Wash' (Osmond, 1988, p. 12). To the south of this line are the standard regions of the South East, South West and East Anglia, and to the north are the other standard regions of England, together with Scotland and Wales (Figure 1.1).

It is remarkable that, taking the boundary as defined above, Britain, in the late twentieth century, divides into two contrasting parts 'in which economic and social factors show a highly consistent pattern and are mutually reinforcing in their effects' (Martin, 1988, p. 401). In the north there are the decaying and comparatively low-income cities of the Industrial Revolution previously dependent upon coal, textiles, metal industries and mechanical engineering, and in the south there is a concentration of economic and political power in the capital and an array of affluent new towns and country towns geared to the growth of high-technology industry and decentralized services.

Perpetual imbalance

While undoubtedly the north became disproportionately disadvantaged by the depression of the 1930s, most of the northern regions have always been the poorest parts of Britain. Even during the Industrial Revolution and its immediate aftermath – when the staple industries of the north (coal, cotton textiles and shipbuilding) achieved their highest outputs – both the level of unemployment and incomes per capita in the north generally compared unfavourably with those in the south. While employers in many parts of the north were dependent both upon labour-intensive methods of production and unstable export markets, industrial activity in the south was geared more to the home market, and London dominated the nation's mercantile activity, banking and finance. Research has recently shown that the highest rates of unemployment in Victorian and Edwardian Britain were primarily found in northern Britain rather than in the south (Southall, 1988); and it has also been revealed that, whereas the northern regions generally had below-average incomes per capita in, for example, 1859–69, 1879–80 and 1911–12 (with Yorkshire & Humberside, the North, Scotland and Wales having the lowest per-capita incomes in these periods), the South East had

Figure 1.1 The north-south division of Britain

the highest per-capita incomes throughout the late-nineteenth/early twentieth century, with East Anglia (in the period 1859-69 and 1879-80) and the South West (in the period 1911-12) similarly having above-average per-capita incomes (Lee, 1986; Martin, 1988). Clearly, the instability of the industrial north and the comparatively advantageous position of the southern economy were 'established features of the British space economy well before the inter-war period' (Martin, 1988, p. 394). A revaluation of the pound in 1924, however, and the world depression of the early 1930s, hastened the collapse of the industries of the north, left the south largely unscathed and consolidated the north-south divide.

For over twenty-five years after the Second World War, interregional imbalance was, however, a relatively minor economic and political issue. Although the rate of unemployment in the north was – in aggregate – twice as high as in the south throughout most of this period, the UK unemployment rate remained at a very low level (rarely exceeding 3 per cent) in keeping with the ceiling advocated by the *White Paper on Employment Policy* (Ministry of Labour, 1944).

After the 'Long Boom' (of 1951-73), the north-south divide became a more serious cause for concern. It was recognized that increasing disparities in prosperity reflected a fundamental imbalance in the distribution of resources and opportunities, and it was becoming widely acknowledged that the dichotomy between a depressed north and an 'overheated' south created severe problems for the whole economy – in other words, how to avoid stagnation on the one hand, and inflation on the other.

During the 1970s and 1980s, forces were at work that were transforming a large part of the economy (see Martin, 1988). Deindustrialization brought about the closure of inner-city premises and a shift of production to more spacious plant elsewhere, and manufacturing employment (which in absolute terms reached its peak in 1966) declined rapidly – particularly after 1979. Although innovations in microelectronics and information technology expanded job opportunities in high-technology industry, growth in this area was insufficient to compensate for the decrease in traditional manufacturing employment. Likewise, whereas there was an increase in employment in the service sector (particularly in banking, financial and producer services), the extent of 'tertiarization' was too little to offset job losses in manufacturing. To a limited extent, however, the Thatcher administration's attempt to transform what it called a 'dependency culture' into a free-market 'enterprise culture' was successful, and – in part – was reflected in the growing number of business registrations in the 1980s.

All of the above changes – to a greater or lesser extent – were caused or motivated by the restructuring of industrial capitalism on a worldwide scale. In abolishing exchange controls in 1979, the government completely exposed the British economy – and the vulnerable north in particular – to the full impact of foreign competition. A new international division of labour had emerged as a result of 'the concentration and internationalization of capital, the closure of the technological gap between first and third worlds, [and] the resulting mass movement of manufacturing to the newly industrializing world' (TCPA, 1987, p. 5), and it was clear that Britain could no longer sustain its manufacturing base at the level at which it had been during the 'Long Boom'.

It was also evident that restructuring was increasing the severity of the 'north-south' problem. Spatially, there was undoubtedly a 'clear inverse relation between deindustrialization and tertiarization' (Martin, 1988, p. 396) – with the north suffering most from the contraction of manufacturing industry, gaining least from the development of high-technology and service activity, and increasingly suffering a residual role in the British economy. Whereas the north had become increasingly dependent upon branch plants and therefore severely disadvantaged by their widespread closure during the recession of 1979–82 (see Fothergill, 1988), the south had dramatically consolidated its position as the British economy's centre of gravity in the 1980s. The South East, for example, contained a heavy concentration of economic activity (both in the private and public sectors), a concentration of venture capital, a concentration of cultural life and a concentration of political power – the governing Conservative Party drawing a large proportion of its electoral and financial support from the region.

The macroeconomics of Thatcherism

Economic transformation at the regional level can only be fully examined in the light of dramatic changes in the macroeconomy. Whereas Conservative governments – in the periods 1951–64 and 1970–4 – applied Keynesian demand-management policy, accepted the need for a mixed economy with a sizeable public sector and broadly demonstrated their belief in the idea of 'one nation', successive Thatcher administrations (after 1979) attempted to use monetarism as a means of controlling the macroeconomy, rolled back the frontiers of State intervention and eschewed the notion of a spatially balanced economy.

Since the Conservative government's principal (or only) macroeconomic objective after 1979 was the control of inflation, and because it believed that price escalation was primarily a result of excessive increases in the money supply, it assumed – in its early years – that the reduction in the money supply was the only route to follow. However, although £M3 (then the principal money-supply indicator) decreased by 5 per cent in 1979–80, it increased by 4.4, 2.3 and 7 per cent in each successive year to 1982–3 – well ahead of Treasury targets. Retail prices (soaring by 21 per cent in 1979–80 and remaining in double figures until 1982) clearly imposed a considerable strain on the money supply. To deflate demand and prices, base rate was increased to 17 per cent in November 1979, which not only directly imposed extra costs on producers but also – through raising the external value of the pound to \$2.29 by October 1980 – further reduced the competitive ability of exporters. The government consequently managed to create the most severe recession in recorded history (worse even than in the 1930s). Instead of reflating the economy, the government adopted 'inverted Keynesianism'. In 1981, in the middle of a slump, tighter fiscal measures were introduced to lower the public-sector deficit – depressing the economy even more.

In an attempt, therefore, to 'squeeze inflation out of the system' and to 'make British industry leaner and fitter', the first Thatcher government, between 1979 and 1981, was instrumental in bringing about a 4-per-cent reduction in the gross domestic product (GDP), a 14-per-cent decrease in manufacturing output (in volume terms) and an increase in unemployment from 1 million to 2.5 million.

Although unemployment continued to rise until it peaked at 3.2 million in June 1986, output began to recover after 1982 - aided by lower interest rates (the average base rate having been reduced to 9.68 per cent in 1984) and by cuts in direct taxation, while export industries benefited from a controlled depreciation of sterling. With a growth of 3.8 per cent in the GDP in 1988 (investment having only just returned to its 1979 level), unemployment fell to below 2 million in January 1989 for the first time since 1980. However, taking the period 1979–88 as a whole, the average annual rate of growth of the GDP was only 2 per cent, the average inflation rate was 8.4 per cent and unemployment averaged 9.5 per cent: a very poor performance overall when compared to that of any other period of government - Conservative or Labour - since the Second World War. By mid-1989, the rate of inflation had increased to an annual rate of 8 per cent, base rate had soared to 14 per cent, unemployment was still at twice the level of what it was in 1979 and the UK faced the likelihood of a £20 billion balance of payments current-account deficit. The third Thatcher government, more than half way through its term in office, appeared to be dangerously close to a further economic crisis and the need to disinflate demand.

While macroeconomic policy had a dramatic effect on growth and employment throughout the 1980s, its impact on the regions was very uneven. High base rates, overvalued sterling and cuts in key areas of public expenditure had a particularly adverse effect on the fragile economy of the north. Whereas there was a modest increase in employment in the south in the 1980s (notably in the service sector), there was a substantial loss of jobs (particularly in manufacturing) in the north - with the rate of unemployment in the North West, the North, Wales, Scotland and Northern Ireland remaining at twice the level of the South East throughout the decade, and falling more slowly after 1986. GDP per capita and average gross weekly earnings were higher in the South East than elsewhere, and so too was average expenditure per household; and, since there were greater job opportunities and greater prosperity in the south, there was a high volume of north–south migration throughout the 1980s - despite house prices in the south rising far more rapidly, and the stock of council houses diminishing more quickly, than in the north. Macroeconomic policy also had an uneven effect on growth within regions. Tight monetary and fiscal measures were, in part, responsible for further inner-city decline, particularly in respect of housing, social services and public transport, while in suburban and rural areas inadequate infrastructure investment and the demise of regional planning exacerbated problems of congestion and 'overheating' - notably in the South East. Although most of these interregional and intra-regional disparities were by no means new, it was evident that the economic and social gaps between the north and south, and between the inner city and the country town, widened substantially during the periods of Thatcher government in the 1980s.

2

THE NORTH-SOUTH DIVIDE - REALITY?

While there are very marked economic and social divisions throughout Great Britain in the late twentieth century (in terms of job opportunities, income distribution, consumer expenditure, house values and health), 'it is demonstrably the case that the concentration of relative socio-economic deprivation and disadvantage is significantly greater in the "north" than it is in the "south" ' Martin, 1987, p. 573). The economic advantages of the south, however, are not only relative but also, in many respects, absolute. In a memorandum to local planning authorities in the South East, the Department of Trade and Industry (1984) set out the economic strengths of the region. According to the department, the South East has the highest concentration of multinational company headquarters in Britain, the greatest number of research and development establishments, an industrial environment that is favourable to enterprise and the highest concentration of business services and invisible export earners. The region, moreover, is part of the European Community's 'Golden Triangle' and is consequently the most favoured alternative location to Continental Europe for inward investment. The north, in contrast, is essentially a branch-plant economy, has comparatively few research establishments, has an industrial milieu often unfavourable to enterprise, has the lowest concentration of business services and invisible export earners and is unfavourably located to exploit trading opportunities on the European mainland.

This chapter examines a comprehensive range of economic and social indicators (both at a regional and urban level) and suggests that the north does worse, on average, than the south in most respects, and that there has been a serious widening in the north-south divide in recent years.

Employment - the great divide

The application of Keynesian demand-management policies during the 'Long Boom' years (approximately 1951-73) ensured that unemployment in the UK remained broadly within the range 1-4 per cent. Reflationary policies (and particularly increased public spending) could always be relied upon to lower the rate of unemployment and to stimulate growth – in certain regions and at times of mild recession. Conversely, disinflationary policies were used to constrain growth and to alleviate labour shortages (in, for example, the South East) in periods of 'unacceptable' inflation. Aided by an expansion of world trade, Keynesian policy enabled manufacturing industry (relatively more important in the north than in the south) to grow without serious disruptions in demand, to keep pace with technological development and – by 1980s' standards – to maintain full employment generally. Throughout much of the 'Long Boom', regional policy (Keynesian in form) had a significant effect on employment in the north. Helped by over-full employment in the South East and Midlands, and an adequate supply of mobile capital, 'carrot-and-stick' measures created or saved 240,000 manufacturing jobs in the Assisted Areas – partly compensating for the contraction of the coal industry and other basic industries (Gudgin and Schofield, 1987). A high level of out-migration away from the northern regions (in net terms, 504,000 people migrated from the north to the south, 1965-73) also helped to maintain relatively full employment in the north (*ibid.*). Increasingly, a national economy was superseding a network of regional economies. Multinational and multi-regional companies were replacing regionally and locally based firms and, with the expansion of the public sector, uniform wage rates became more and more the norm. Clearly, therefore, during the 'Long Boom', there was a very marked interregional convergence in employment opportunities.

By 1981, however, substantial changes in employment became evident. Over the previous ten-year period, both in the northern and southern regions, there was a serious reduction in employment in manufacturing industry: by 659,605 (or 27.77 per cent) in the north and by 1,653,360 (or 23.92 per cent) in the south (June 1971 to September 1981). Conversely, there was a large increase in service employment in both areas: by 323,854 (or 11.28 per cent) in the north and by 462,343 (or 10.05 per cent) in the south over the same period. It is of note that while percentage changes by economic sector were not dissimilar, because of its much larger employment base, the south experienced greater employment losses and gains in absolute terms than the north (Breheny, Hall and Hart, 1987). During the 1970s, regional policy was having a diminished impact on job creation in the Assisted Areas (Keeble, 1980b; 1987). Whereas 121,000 jobs (net) were created or saved in manufacturing industry in these areas in the period 1971-6, only 73,000 jobs were generated in the years 1979-81 – a result of recession and the closing down of old plant (Moore, Rhodes and Tyler, 1986), together with the impact of cuts in regional aid and the abandonment of associated incentives (see Chapter 3).

During the 1980s, stark regional divergencies became apparent. This was associated with the continuing decrease in manufacturing employment and coal-mining, and the further growth of services. In 1980-1, manufacturing output fell by 17 per cent and – to cut wage costs – employment in manufacturing was reduced by as much as 25 per cent and has not yet recovered (Gudgin and Schofield, 1987). From the mid-1980s, a large closure programme substantially reduced employment in the coal-mining industry - the number of jobs in coal having already been cut by the reduced demand for coal from the Central Electricity Generating Board, the steel industry and the economy in general. Private service industries alone accounted for all the employment generated in the 1980s, but with the south benefiting significantly more than the north (*ibid.*).

Whereas in the 1930s the main north-south structural difference in manufacturing industry was between a concentration on coal, steel, ships and textiles in the north and the development of a wide range of consumer-goods industries in the south, in the 1980s (while the north was still significantly dependent on its old and vulnerable basic industries) the south benefited from the growth of research and development (R & D), the expansion of high-technology industry and new-firm formation - even during the recession of 1980-2 (Townsend, 1983; Gould and Keeble, 1983; Keeble, 1986, 1987). It was very clear that to the end of the century and beyond, 'industries [were] ... not going to be born in yesterday's regions... . Britain's future, if it has one, is in that broad belt that runs from Oxford and Winchester through the Thames Valley and Milton Keynes to Cambridge' (Hall, 1981, p. 536). Keeble and Gould (1986) and Keeble and Kelly (1986) believed that this pattern of growth would emerge because of the residential attractiveness of the South East, while Marsh (1983) and Segal Quince and Partners (1985) suggested that links with and spin-offs from universities and public research centres would have a catalytic effect on industrial development. Breheny, Cheshire and Langridge (1983) pointed out that the development of high-technology industry adjacent to the M4 had been attributable to spin-offs from nearby and long-established government research institutions, as well as to proximity to the capital (with its centres of decision-making, information and communication). There was clearly an increasing concentration of new and growing industry in the South East, in large part a result of cumulative causation (where external economies in a 'core' area widen the degree of regional disparity). Tyler (1987) showed that by using location quotients (LQs) it was possible to indicate the degree to which different counties (or regions) had attracted new and growing industry in comparison to Great Britain overall. Whereas the most successful counties were in the south (i.e. Hertfordshire, Berkshire, Surrey, Hampshire and Gloucestershire with LQs ranging from 249 to 162), the least successful areas were all in the north (notably Powys, Dyfed, Grampian, South Yorkshire and Orkney, with very low LQs ranging from 36 to as little as 8). Clearly, LQs could very easily be used to measure the extent of interregional disparity in most forms of employment.

By the 1987 general election, it was being increasingly claimed that Thatcherism had widened the north-side divide (*The Financial Times*, 1987a, 1987b; House of Commons, 1987; *The Independent*, 1987a, 1987b; *The Sunday Times*, 1987a, 1987b). The Labour Party and the Alliance gave the issue prominence in their manifestos but Mrs Thatcher 'consistently dismissed the idea of any serious "two nation problem" - despite evidence to the contrary published by her own government departments' (Martin, 1987, p. 571). The evidence comprised the unemployment projections that had been included in the government's submission of it's regional development programme to the European

Regional Development Fund (Department of Trade and Industry, 1986), and the findings of the *Census of Employment*, 1986 (Department of Employment, 1987a). The *Census of Employment* revealed that whereas total employment in Great Britain had decreased by 745,000 in the period between 1979 and mid-1986, in the north the number of employed had fallen by as much as 1,101,000 - offset marginally by an increase of 356,000 jobs in the south (Table 2.1). Figure 2.1 shows that (with the exception of the East Midlands) all the northern regions, and particularly the North West, experienced substantial reductions in employment over the first seven years of Thatcherism, while in each of the southern regions employment increased, albeit negligibly in the South East.

The *Census of Employment* also showed that in the period 1979-86, employment in manufacturing declined by almost 2 million (a reduction of 28 per cent - more than double that of any other industrial nation), while employment in services expanded by 860,000. However, whereas three-quarters of manufacturing job losses were in the north, two-thirds of the increase in service employment was in the south. Although a total of 1 million jobs were created in the comparatively short period 1983-6, this number must be contrasted with the 1.75 million jobs lost (1979-83), and it is notable that less than a quarter of these new jobs were in the north (Martin, 1986). The overall expansion in employment during economic recovery after 1983 was not unlike the upswing in the 1970s, except that it was much more favourable to the service sector and to the south. While employment disparities widen during recession, they normally narrow during recovery. This did not happen in the middle or late 1980s (Armstrong and Riley, 1987). MacInnes (1988) showed, moreover, that (contrary to Conservative claims) Britain's employment performance over time did not improve at all. Employment in Great Britain as a whole increased at the same rate under the Conservatives (1.9 per cent, 1982-7) as under Labour (1975-9), but in contrast to Labour's

Table 2.1 Trends in total employment, Great Britain, 1979-86

| | Employment change | | | |
| | 1979-86 | | 1983-6 | |
	('000)	(%)	('000)	(%)
South East	+ 172	+ 2	+ 484	+ 6
East Anglia	+ 101	+ 13	+ 100	+ 13
South West	+ 83	+ 5	+ 99	+ 6
East Midlands	+ 6	—	+ 94	+ 6
Yorkshire & Humberside	- 130	- 6	+ 74	+ 4
North	- 138	- 10	+ 39	+ 3
Wales	- 145	- 13	- 2	—
Scotland	- 172	- 8	+ 12	+ 1
West Midlands	- 173	- 7	+ 90	+ 4
North West	- 349	- 12	+ 15	+ 1
Great Britain	- 745	- 3	+ 1,005	+ 4
'South'	+ 356		+ 777	
'North'	- 1,101		+ 228	

(*Source:* Derived from Department of Employment, *Census of Employment*, 1987a.)

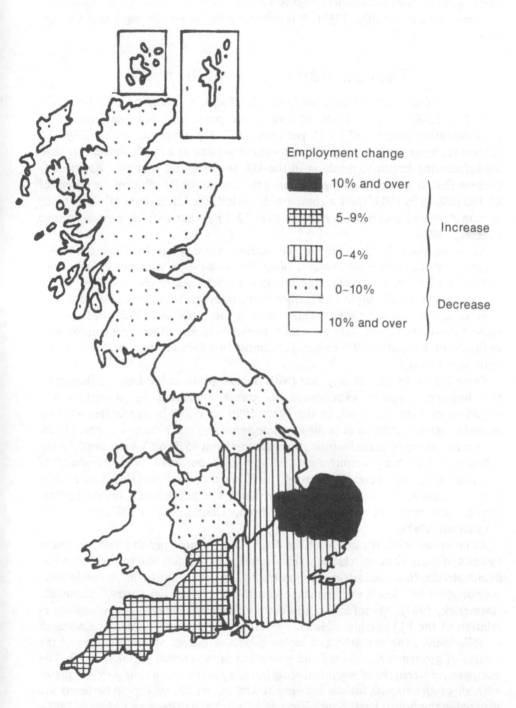

Employment change

■ 10% and over

▦ 5-9% } Increase

▥ 0-4%

⦙ 0-10% } Decrease

▢ 10% and over

Figure 2.1   Employment change in Great Britain, 1979–86 (Source: Department of Employment)

record on employment, the Thatcher government progressively abandoned manufacturing and increasingly aligned itself particularly to high technology and financial services (Martin, 1987). It is to these areas of employment that we now turn.

The deindustrialization of the north

Over 3.1 million workers were shed from manufacturing industry in the UK in the period 1966–83 – a decrease of nearly 40 per cent, and the volume of manufacturing output fell by 18 per cent (1973–83). Over the period 1970–83, moreover, manufacturing exports increased in volume at a much slower rate than manufacturing imports, resulting in the UK incurring its first-ever balance-of-trade deficit in manufactured goods since the Industrial Revolution – a deficit of £2,100 million in 1983. Over a comparable period, the UK's share of world trade in manufactured goods slumped from over 12.5 per cent to 9 per cent in volume (Keeble, 1987).

At a regional level, it is in the north rather than in the south that deindustrialization is most evident. Since the mid-1960s, and particularly since 1979, there has been a dramatic decline in the traditional manufacturing base of the north in terms of output and employment, whereas, in contrast, the decline of Greater London as a manufacturing centre in the 1960s and 1970s subsequently slowed down; and there were noticeable increases in manufacturing employment in the South East after 1979 (where for almost two decades there had been large-scale job losses).

Table 2.2 shows very clearly that (with the exception of the East Midlands) all the northern regions experienced a greater decrease in manufacturing employment than the south in the years 1979 to 1987. In aggregate, whereas manufacturing employment in the north decreased by more than 32 per cent (with the north's share of manufacturing jobs falling from 65.3 to 62.4 per cent), in the south manufacturing employment fell by 23.3 per cent (and its share of manufacturing jobs increased from 34.7 to 37.6 per cent). It was thus very evident that in terms of manufacturing employment, the north-south divide became increasingly apparent in the 1980s (Owen, Coombes and Gillespie, 1983; Townsend, 1983).

At an urban level, the extent of the divide is even more pronounced. Southern towns and cities contain relatively few declining industries, and in the process of deindustrialization industries concentrated in the south have undergone proportionately less employment contraction than industry nationally (Damesick, 1987). Manufacturing in the south clearly benefits from locating in relation to the M3 corridor, Heathrow and Gatwick airports, defence research establishments, the scientific and technological resources of Cambridge and the centres of government, finance and education in the capital (Cooke, 1987). The comparative strengths of manufacturing industry in the south thus partly explains why all of the top-20 fastest growing towns in Britain (except Aberdeen) are situated in the South East, South West and East Anglia (Begg and Moore, 1987).

Regional disparities in the availability and rents of industrial land are a further

Table 2.2   Decrease in manufacturing employment, UK, 1979-87

| | Employment | | | | Change |
| | 1979 | | 1987 | | 1979–87 |
	('000*)	(% UK)	('000*)	(% UK)	(%)
East Anglia	206.4	(2.8)	205.1	(4.0)	−0.6
South West	441.0	(6.1)	367.8	(7.1)	−16.6
East Midlands	606.5	(8.3)	492.0	(9.6)	−18.9
South East	1,875.7	(25.8)	1,363.0	(26.5)	−27.3
West Midlands	986.0	(13.6)	698.4	(13.6)	−29.2
Northern Ireland	144.8	(2.0)	100.4	(1.9)	−30.7
Wales	315.1	(4.3)	26.3	(4.0)	−34.5
Scotland	605.4	(8.3)	394.2	(7.6)	−34.9
North	410.6	(5.6)	263.8	(5.1)	−35.8
Yorkshire & Humberside	707.9	(9.7)	448.7	(8.2)	−36.6
North West	971.3	(13.4)	610.7	(11.9)	−37.1
UK	7,271.2	(100.0)	5,145.2	(100.0)	−29.2
'South'	2,523.1	(34.7)	1,935.9	(37.6)	−23.3
'North'	4,748.1	(65.3)	3,209.3	(62.4)	−32.4

Note
* Numbers rounded to nearest thousand.

(*Source:* Department of Employment.)

manifestation of uneven deindustrialization. Fothergill, Kitson and Monk (1987) revealed that, at the end of the 1979–82 recession, there were respectively 15, 13 and 11.3 hectares of available industrial land per 1,000 employees in Wales, Scotland and the North, but only 7 hectares per 1,000 employees in the South West, 6.9 in East Anglia and 2 in the South East: a reflection principally of a comparatively low level of derived demand in the northern regions. In the subsequent period, 1982–6, broadly the same variations in demand resulted in industrial rents in Wales and the North falling on average by 0.5 and 0.3 per cent per annum and standing still in Scotland, whereas in the South West, South East and East Anglia average annual rental increases amounted to 5.5, 4.5 and 0.5 per cent respectively. By 1986 industrial rental values ranged from as little as £11.8 per m² in the North to as much as £28.06 per m² in the South East (or £35.77 per m² in London) (Hillier Parker, 1987).

High-technology industry

The regional pattern of high-technology industry in Britain has been the subject of numerous studies in recent years (see Begg and Cameron, 1988; Hall *et al.*, 1987), and each indicated that high-technology industry is overwhelmingly located in the south, with both manufacturing and services particularly concentrated in the South East including Greater London. Some regions, however, have negligible shares of high-technology activity, for example, Yorkshire & Humberside and Wales (Table 2.3). It is of note that high-technology industries are generally distributed in inverse proportion to the distribution of 'traditional' industry (Begg and Cameron, 1988).

Table 2.3 Regional distribution of high-technology employment in Great Britain, 1981–4

| | High technology | | | | All high technology | | Share of GB high technology (%) |
| | Manufacturing | | Services | | | | |
	1981*	1984*	1981*	1984*	1981*	1984*	1984
South East	118.9	116.1	153.6	150.8	129.9	127.9	44.3
Greater London	77.0	70.9	162.5	160.2	104.1	101.2	16.8
Rest of South East	159.6	157.9	145.0	142.2	155.0	152.6	27.5
South West	121.0	129.8	102.0	97.1	115.0	118.7	8.7
East Anglia	60.5	86.7	120.0	145.9	79.4	106.2	3.6
North West	113.8	113.0	75.5	67.0	101.0	97.4	10.7
East Midlands	110.2	98.3	54.3	59.7	92.5	85.2	6.0
West Midlands	98.6	87.7	75.5	78.1	91.3	84.4	7.9
North	83.1	96.3	51.6	56.5	73.1	82.8	4.2
Scotland	76.2	83.5	75.3	68.4	75.9	78.4	7.2
Wales	70.2	78.2	55.0	66.5	65.3	74.2	2.9
Yorkshire & Humberside	47.5	42.9	55.3	54.0	50.0	46.7	3.4
Great Britain	100.0	100.0	100.0	100.0	100.0	100.0	100.0

Note
* Location quotients, GB = 100.
(Source: Begg and Cameron, 1988.)

Within the South East, the principal growth areas for high-technology industry are along the M3/M4 corridors and around Cambridge. A major locating influence in, for example, the production of computer software is the demand generated by head offices in London (Hall *et al.*, 1987), while the South East has for long been the dominant location for R & D. In 1976 the region contained 57 per cent of all R & D jobs - government research establishments in close proximity to the capital accounting for a significant proportion of this figure (Howells, 1984). While high-technology industries are by no means absent from Wales and Scotland (the development of 'Silicon Glen' in the 1960s and 1970s being a major growth achievement), they nevertheless conform to the inherently vulnerable branch-plant pattern that generally characterizes industry in much of the north (Randall, 1987).

Interregional variations in the demand for sites for high-technology industry largely account for values falling from, for example, £2,471,000 per hectare in Maidenhead or £1,730,000 per hectare in Crawley to as little as £222,000 and £148,000 per hectare in Leeds and Cardiff in 1986 (Organ, 1987).

At an urban level, not only are there striking interregional disparities in the distribution of high-technology industry, but there is also a very clear inverse relationship between high-technology and traditional employment. Using location quotients (LQs), Begg and Cameron (1988) showed that (with the exception of Newton Aycliffe (Co. Durham) the top-five urban locations for high-technology employment were all in the south (Stevenage being the most favoured location with a LQ of 498.4), whereas the bottom-five locations were all in the north (Peterlee having a LQ of only 2.1). In aggregate (and taking into account both high-technology manufacturing and high-technology services), LQs were considerably higher in the south, in contrast to traditional employment where LQs were higher in the north (Table 2.3).

Services

In absolute terms, service employment is highly concentrated in the South East. As Table 2.4 shows, over 5 million service jobs were located in this region (more than 37 per cent of the national total) in both 1979 and 1987. The south as a whole, moreover, increased its share of service employment from 47.5 to 48.7 per cent of the national total in the period 1979-87, whereas the north decreased its share from 52.5 to 51.3 per cent. Two northern regions, the North West and Wales, moreover, suffered an absolute decline in service employment (1979-87). Thus, even in the principal growth sector of the economy, parts of the north were not immune from substantial job losses (Breheny, Hall and Hart, 1987). It is clear, therefore, that those regions that had shown the least decline in manufacturing were broadly the same as those that had experienced the greatest rate of growth in service employment, whereas those that had suffered the collapse of manufacturing found it difficult to attract or even retain jobs in service industries.

Table 2.4 Increase in service employment, UK, 1979-87

	Employment				Change
	1979		1987		1979-87
	('000*)	(% UK)	('000*)	(% UK)	(%)
East Anglia	401.5	(3.0)	509.6	(3.4)	+26.9
East Midlands	752.6	(5.5)	881.7	(5.9)	+17.1
West Midlands	1,071.2	(7.9)	1,199.8	(8.1)	+12.0
South East	5,044.3	(37.2)	5,623.2	(37.9)	+11.5
Yorkshire & Humberside	1,045.7	(7.7)	1,165.9	(7.8)	+11.5
South West	995.5	(7.3)	1,093.7	(7.4)	+9.9
North	663.9	(4.9)	706.3	(4.8)	+6.4
Northern Ireland	317.6	(2.3)	332.2	(2.2)	+4.6
Scotland	1,223.4	(9.0)	1,253.6	(8.4)	+3.3
North West	1,487.9	(11.0)	1,477.1	(9.9)	-0.7
Wales	571.2	(4.2)	561.8	(3.8)	-1.6
UK	13,570.0	(100.0)	14,847.2	(100.0)	+9.4
'South'	6,441.3	(47.5)	7,226.5	(48.7)	+12.2
'North'	7,128.7	(52.5)	7,620.7	(51.3)	+6.9

Note
* Numbers rounded to nearest thousand.

(*Source:* Department of Employment.)

In discussing interregional patterns of service employment, it is necessary to disaggregate the sector into the following three broad categories: distribution, hotels and catering; business and financial services; and public administration (see Table 2.5). First, in examining employment in the distribution, hotel and catering industries, not only is it evident that employment is greatest in absolute terms in the South East, but that it increased faster in this region than in the UK as a whole in the period 1979-87. Second, in considering employment in the financial services, the locational dominance of the South East is very apparent - the region containing about half the total number of jobs in this category in both 1979 and 1987. The concentration of industrial power in the South East necessitates access to a very wide range of business and financial services not provided 'in-house', whereas in the north a preponderance of branch firms and a lower level of demand justify far fewer outside services - a situation exacerbated by the growing annexation of northern accountancy, banking, law, advertising, stockbroking and insurance firms by organizations based in London. A substantial proportion of business and financial services in the north are themselves provided by establishments chiefly based in the capital: over 70 per cent according to a survey undertaken in three provincial regions in the late 1980s (TCPA, 1987). That employment in business and financial services in the South East barely increased at the national rate, in the period 1979-87, is therefore of limited significance. Third, with regard to public administration, employment is again concentrated in the South East but not to the same degree as business and financial services. In the period 1979-87, employment growth in the South East lagged behind the national rate of increase, while employment in most northern regions expanded at a more than proportionate rate, suggesting a continual decentralization of jobs from London.

The centralization of capital and enterprise

An examination of interregional patterns of employment clearly suggests that the north is at a considerable disadvantage in attracting or retaining jobs in both the manufacturing and service sectors of the economy. However, the distribution of employment, *per se*, does not explain why growth sectors (such as high-technology industry and business and financial services) and new-company formation rates are concentrated in the south. While the expansion of new industries in the north may have been disadvantaged by high transport costs to markets earlier in the century, in the 1980s transport costs normally account for a very small percentage of an industry's total costs and, therefore, can be generally discounted as a locational influence.

It can be proposed, however, that the principal reason for the decline of economic activity in the north is the trend towards national and multinational corporations and their desire to concentrate in the South East. It was revealed that 485 out of 1,000 of the biggest companies in the UK had their headquarters in London, and that in one South East county alone, Berkshire, 28 firms had their headquarters, compared with 15 in the whole of the Northern region (Fothergill and Vincent, 1985). It was also reported that only 10 per cent of the 100 largest companies in the UK had their head offices in the north (Watts, 1988): Birmingham, Glasgow, Leeds and Manchester had but one head office each, while Liverpool, Newcastle and the whole of Wales had none (Hughes, 1988b). The 100 largest companies, moreover, controlled at least half of industry's output by the early 1980s (compared with less than one-fifth in the 1920s) - a result of merger mania in the late 1960s and early 1970s (Regional Studies Association, 1983).

Southern-based organizations were thus absorbing an increasing proportion of northern firms. In 1983 to 1987, two-thirds of the public companies based in the North region were absorbed by national or multinational corporations, while 80 per cent of the region's workforce were employed in branch plants by the middle of the decade (TCPA, 1987). A similar situation had occurred in Scotland where taken-over companies tended to lose most of their operations (including key management functions) to the south, with only production management and routine clerical and manual work remaining 'north of the border'. It was ironic that whereas in the past it was the smaller and weaker firms that were subject to takeover, in the 1980s (and not least in Scotland) it was the larger and faster-growing companies that were being absorbed (Henry, 1987). A particularly disturbing aspect of the emergent branch economy of the north was that while in London there was a ratio of 100 white-collared workers to 54 manual workers in the late 1980s, in Cleveland, Merseyside and South Yorkshire the ratio of white-collared workers to manual workers was 100 to 110 (TCPA, 1987). The transfer of managerial and professional jobs to the south has thus produced a distorted employment profile in the north with undoubted reverse multiplier effects on the northern economy.

Table 2.5   Employment change in service industries, UK, 1979–87

	Distribution, hotels and catering, repairs			
	1979 ('000)	1987 ('000)	1987 M (%)	1987 F (%)	Change 1979–87 (%)
South East	1,442.3	1,549.2	18.7	23.3	7.4
East Midlands	256.6	291.8	15.6	23.5	13.7
West Midlands	363.0	373.3	15.4	21.8	2.8
Yorkshire & Humberside	360.1	389.2	15.6	28.1	8.1
East Anglia	134.8	163.0	16.8	25.5	21.0
South West	357.9	358.2	18.7	27.0	0.1
North	227.1	201.7	13.2	25.3	− 11.2
Scotland	391.0	382.9	15.6	25.5	− 2.1
Northern Ireland	78.9	81.5	15.1	18.8	3.3
Wales	170.4	174.3	15.2	26.3	2.3
North West	476.3	493.1	16.5	27.6	3.5
UK	4,260.9	4,469.4	16.9	24.7	4.9

Note
M Male; F Female.
(*Source*: Department of Employment.)

The greater 'face-to-face' accessibility of southern-based giant firms to sources of finance in the City of London might be the principal reason why head offices are seriously under-represented in the north. Financial institutions, largely with their headquarters in London, tend to be 'regionist' in their investment policy; for example, in 1985, 60 per cent of venture capital went to firms in the South East (Mason, 1987). The image of the north, often unflatteringly purveyed by the media, cannot be discounted as having little significant effect on investors.

In addition to investment considerations, locational factors play a major part in concentrating many of the functions of the major corporations in the south. The advantages to high-technology industry and offices of a location in, for example, the M4 corridor have been set out above. Similarly, the low adoption of information technology in the peripheral regions puts much of the north at a disadvantage. Insurance companies, for example, might find it increasingly impracticable to retain regional offices in the north, and national newspaper publishers may be unable to exploit modern technology fully in their printing operations. New developments in information technology could be confined to the south in the future, and further reinforce existing patterns of employment (Goddard, 1985).

The centralization of capital in the south is having a negative effect on the ability of the north to regenerate itself through its own enterprise (Regional Studies Association, 1983; Damesick, 1987; Mason and Harrison, 1989). Table 2.6 shows, for example, that not only were there far fewer business registrations in each of the northern regions than in the South East but that the net gain in registrations in the north (both in absolute and percentage terms) was considerably less than the net gain in the south. Mason and Harrison (1989) suggested that investment had been disproportionately attracted to firms in the

Table 2.5 *Cont.*

Banking, finance, insurance, business services and leasing					Public administration and other services				
1979	1987	1987		Change	1979	1987	1987		Change
		M	F	1979–87			M	F	1979–87
('000)	('000)	(%)	(%)	(%)	('000)	('000)	(%)	(%)	(%)
807.1	1,139.4	14.8	15.9	41.2	2,182.1	2,361.0	22.4	42.7	8.2
71.5	93.2	5.2	7.2	30.3	349.9	417.1	18.4	38.5	19.2
114.2	180.5	7.5	10.5	58.1	497.5	557.8	17.1	40.1	12.1
96.5	146.0	8.0	8.2	51.3	476.6	526.2	18.8	41.8	10.4
42.1	70.0	7.9	10.1	66.3	181.8	212.2	17.7	38.8	16.7
106.9	162.4	9.0	11.6	51.9	444.2	488.7	21.0	42.2	10.0
57.4	77.4	6.5	7.7	34.8	314.5	370.6	22.7	47.9	17.8
124.0	164.1	7.9	9.6	32.3	573.8	622.4	22.1	45.3	8.5
24.9	28.1	5.3	6.3	12.9	192.5	204.2	33.9	54.8	6.1
44.4	63.9	6.9	8.1	43.9	290.5	281.3	22.5	45.0	– 3.2
157.9	199.1	7.7	10.0	26.1	685.1	656.0	18.9	40.3	– 4.2
1,644.1	2,332.8	9.9	11.6	41.9	6,182.9	6,693.2	20.9	42.5	8.3

South East under the Business Expansion Scheme (BES) (see Chapter 3). In their view, this was because the small-business sector of this region was more dynamic than in the rest of the country; BES fund managers were concentrated in London and for reasons of awareness, logistics and opportunity tended to favour investment in the South East; and because prospectus share-issues were sponsored almost entirely by London-based financial intermediaries within the fields of low-risk property and consumer-oriented services located

Table 2.6 Business registrations, Great Britain, 1979–87

	Stock		Net gain	
	1979	1987	1979–87	
	('000)	('000)	('000)	(%)
South East	423.7	525.0	101.4	23.9
East Anglia	52.2	62.2	10.1	19.3
South West	124.0	146.6	22.6	18.2
East Midlands	88.2	103.7	15.6	17.7
West Midlands	111.8	129.1	17.3	15.5
Scotland	93.4	106.8	13.4	14.3
Wales	70.4	78.7	8.3	11.8
North	52.4	58.5	6.2	11.8
Yorkshire & Humberside	102.0	113.9	11.9	11.7
North West	128.3	138.7	10.5	8.2
Great Britain	1,246.2	1,463.4	217.2	17.4
'South'	599.9	733.8	133.9	22.3
'North'	646.3	729.6	83.3	12.9

(*Source:* Central Statistical Office, *Regional Trends*.)

overwhelmingly in the South East. It is also of significance that northern regions have a high percentage of manual and clerical labour, and thus a low propensity to develop new products and to establish new businesses (Gould and Keeble, 1983; Whittington, 1983). It is notable that in the north there has been a traditional reliance on large plants with a majority of jobs within manual categories (Gudgin and Fothergill, 1984), but even if an enterprise culture re-emerged in the north, the lack of finance would be a major constraint on the development of new firms (TCPA, 1987).

Employment in the 1990s

It seems highly probable that, if left mainly to the market, much of the growth of jobs in the early 1990s will be in the south, thereby reinforcing regional inequalities. Tyler and Rhodes (1986) forecast that there would be a net increase of 900,000 jobs in Great Britain in the period 1985–95, but that 417,000 of this number would be in the South East (Greater London taking a sizeable share). In the South East (according to their projection), the absolute decline in manufacturing will level off, and the region will continue to contain a large proportion of the fast-growing high-technology and commercial sectors, and accommodate a substantial concentration of company headquarters. Gudgin and Schofield (1987), furthermore, predicted that over a longer-term period, stark regional disparities will remain at least until the year 2000.

Inadequate infrastructure development in the north, particularly in the case of air and rail transport, will ensure that the South East continues to maintain or increase its share of employment. Whereas London's Heathrow and Gatwick airports had throughputs of 37.5 and 20.8 million passengers in 1988, Manchester, Newcastle, Glasgow, Edinburgh and Aberdeen airports handled a total of only 18.7 million passengers in the same year. More importantly, while the two London airports offer business travellers 'an unrivalled range of direct international services ... by contrast, airports outside the south play a subservient role, acting essentially as feeders to London's airports and offering few direct international services' (TCPA, 1987, p. 18). Clearly, the lack of proximity to a major international airport is a major deterrent to the location of new employment opportunities in high-technology industry and services geared to foreign trade, and in this respect the north is at a considerable disadvantage.

The further expansion of employment in the South East, however, will (unlike the past) be in the eastern rather than western counties of the region – due to the development of London's Docklands, Stansted airport and the Channel tunnel. In respect of the Channel tunnel, it was forecast that the Midlands rather than the South East would benefit most from the creation of new jobs in construction, with gains of 3,620 jobs by 1989 compared to only 2,710 in Kent and a trifling 15 in the rest of the South East (Channel Tunnel Group, 1985). But with its motorway links via the M20 and M25 to the M3/M4 and M11 corridors (Figure 2.2), the tunnel in the longer term will undoubtedly attract fast-growing manufacturing industries to the south-eastern sector of the South East, where accessibility to both international and domestic markets might (in aggregate) be maximized. Whereas Vickerman (1987, p. 191) suggested 'that such industries do

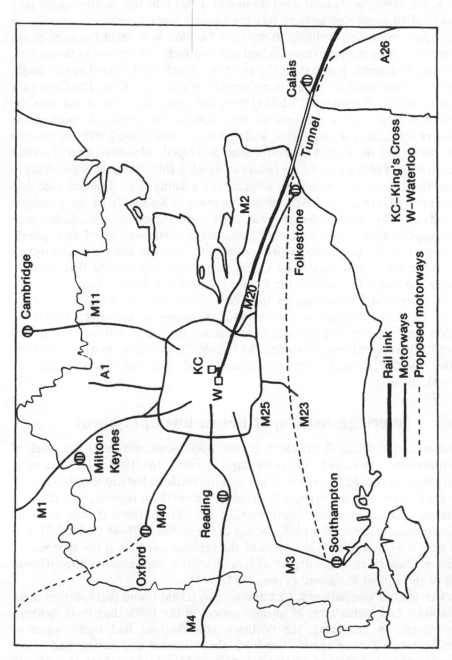

Figure 2.2 Motorways and the Channel link

not have a great effect on employment in total', the Channel Tunnel Joint Consultative Committee (1987) predicted that 14,000 new jobs could be created in Kent in the 1990s, set against a net decline of 3,000 jobs lost in diminished port activity. High-speed rail services (via the tunnel) from London to Brussels or Paris–Lyon, moreover, will help to maintain London as a world financial centre and generate further employment in business and financial services in the capital.

Of major concern, however, is the extent to which the Channel tunnel might have an adverse effect on the economy outside of the South East. The Centre for Local Economic Strategies (CLES) (1989), for example, pointed out that the North had won only 4 per cent of the £700-million worth of construction contracts available in the late 1980s, and that only 1,000 of the 7,900 construction jobs created by 1989 were in the North West and North. However, a more serious concern in the 1990s would be the failure to develop fully Channel-tunnel links to the north, and to rely instead on utilizing the existing West London Line (via Kensington Olympia) for freight or interchanges at King's Cross for passenger connections. This, in the view of Steer Davis and Gleave (1989) would heighten the perception that current industrial bases in the north are 'out of date, poorly connected to other production units in other EC countries and with substandard links for business travellers'; and CLES (1989) similarly warned that, without heavy government investment, the Channel tunnel will damage prospects in the north, exacerbate overheating in the South East and widen the north-south divide. The Confederation of British Industry suggested, moreover, that if the north is not served by an adequate infrastructure, the tunnel will suck investment out of the overheated South East into the Pas-de-Calais with its better connections to the 'Golden Triangle' between Brussels, Munich and Paris (Black and Liniecki, 1989).

Interregional disparities in unemployment

A major and traditional indicator of the north-south divide is the level of unemployment measured in percentage terms. In 1988, for example, unemployment ranged from over 16 per cent in Northern Ireland to as little as 5.1 per cent in East Anglia (Figure 2.3). Among the northern regions, only the East Midlands and (to a much lesser extent) the West Midlands had rates of unemployment lower than the UK average in the period 1979–88 (Table 2.7). It is thus very clear that a 'central feature of the regional problem is the existence of persistent disparities in the degree of labour market imbalance in the different parts of the United Kingdom' (Tyler, 1987, p. 1).

In like-for-like comparisons, Champion *et al.* (1986) found that most northern urban areas had higher rates of unemployment in the 1980s than their southern counterparts; in particular, the northern conurbations had higher rates of unemployment than Greater London, their towns and cities had more unemployment than their equivalent in the Outer Metropolitan Area of the South East, and the free-standing cities of the north had proportionately more of their labour force out of work than the free-standing cities in the south. A similar like-for-like comparison of unemployment rates recognized that, since the majority of

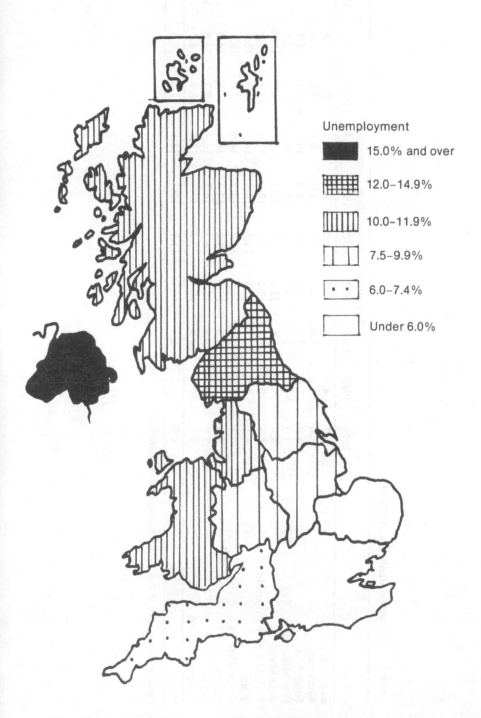

Figure 2.3   Unemployment in the UK, 1988 (Source: Department of Employment)

Table 2.7 Unemployment, UK, 1979–88

					Percentage of labour force unemployed					
	1979	1980	1981	1982	1983	1984	1985	1986	1987	1988
East Anglia	3.7	4.7	7.3	8.5	9.0	8.6	8.7	8.7	7.5	5.1
South East	3.0	3.8	6.4	7.7	8.4	8.4	8.6	8.6	7.5	5.5
South West	4.8	5.6	7.9	9.1	9.8	9.8	10.1	10.1	8.8	6.6
East Midlands	4.0	5.6	8.7	9.9	10.6	10.7	10.7	10.6	9.6	7.6
West Midlands	4.8	6.8	11.6	13.6	14.2	13.7	13.7	13.5	11.9	9.0
Yorkshire & Humberside	5.1	6.8	10.5	12.2	12.8	12.8	13.1	13.5	12.2	9.9
Wales	6.5	8.4	12.2	13.8	14.3	14.4	14.9	14.9	13.0	10.9
North West	6.1	7.8	11.7	13.6	14.7	14.7	14.9	14.9	13.4	11.0
Scotland	6.9	8.5	11.6	13.0	13.8	14.0	14.2	14.6	14.0	11.8
North	7.9	9.9	13.8	15.5	16.3	16.3	16.6	16.3	14.8	12.4
Northern Ireland	9.1	10.9	14.4	16.1	17.2	17.7	17.6	18.6	18.4	16.6
UK	4.9	6.2	9.4	10.9	11.7	11.7	11.8	11.8	10.6	8.1

(*Source:* Department of Employment, *Employment Gazette.*)

the population live and work in large cities despite decentralization (Fothergill and Gudgin, 1982; Martin and Rowthorn, 1986), it is at the urban level that the north-south divide can best be examined (Armstrong and Riley, 1987). Referring to data for 1986, Armstrong and Riley showed that whereas Scotland, the North West and the North had respectively 16.1, 18.1 and 35.4 per cent of their workforce in travel-to-work areas with unemployment in excess of 20 per cent, in East Anglia, the South East and the South West, respectively zero, 0.5 and 0.8 per cent worked in areas with such high unemployment.

Based on Armstrong and Riley's analysis, an examination of unemployment at county level offers a further manifestation of interregional disparity. Table 2.8 reveals that in March 1988 the rate of unemployment in the 6 northern metropolitan counties was higher than in Greater London; unemployment in the 16 principal manufacturing counties in the north was higher than in the 8 principal manufacturing counties in the south; and the proportion of labour force out of work in the 11 largely agricultural counties in the north was greater than the proportion out of work in the 14 mainly agricultural counties of the south.

Tyler (1987) shows that there are also clear disparities in unemployment between the development and intermediate areas of the north, the rest of the country and Great Britain as a whole - as might be expected. In aggregate, in 1986 unemployment in the development areas averaged 19.5 per cent, in the intermediate areas 16.1 per cent of the labour force was out of work, in the rest of the country 10.7 per cent were unemployed and in Great Britain overall the unemployment rate was 13.2 per cent. The unemployment rate in the development areas has, of course, been higher (and sometimes substantially higher) than the national average since these areas were first designated shortly after the Second World War.

Over time, however, regional unemployment converges and diverges from the national rate. From the mid-1960s to the late 1970s, unemployment rates in the

Table 2.8   Like-with-like unemployment comparisons, Great Britain, 1988

Counties	Unemployment 1988 (March) (%)
Greater London	8.30
'Northern' metropolitan counties (6)	14.30
Counties with above-average employment in manufacturing:	
'Southern' (8)	6.00
'Northern' (16)	11.10
Counties with above-average employment in agriculture, forestry and fishing:	
'Southern' (14)	7.96
'Northern' (11)	10.15

Note
Number of counties shown in brackets.

(*Source:* Department of Employment, *Census of Employment*; *Employment Gazette*.)

north and south converged – not least due to a disproportionate increase in the rate of unemployment in the South East and particularly in Greater London from the middle of the 1970s (Manners, 1976; Frost and Spence, 1983). This prompted both Labour and Conservative governments to shift the emphasis of industrial and planning policy from regional support to the regeneration of the inner cities (see Chapter 3). As Figure 2.4 shows, however, divergence was strikingly evident during the recession 1979–81, and remained very apparent throughout the 1980s since, in general, unemployment decreased in the south and increased in the north. Clearly, economic recovery after 1982 had been very shallow and the north, in particular, continued to suffer heavy job-losses in manufacturing throughout the 1980s (Armstrong and Riley, 1987).

The plight of the north was fully documented by the Department of Employment and Department of Trade and Industry (DTI, 1986) in their report to the European Commission in connection with the government's applications for aid from the European Regional Development Fund. The report pointed out that in Northern Ireland, 'a high rate of unemployment [was] ... likely to be the norm for some time'; in the Northern region, 'the situation will not improve until a number of fundamental problems are resolved' (in relation, for example, to the development of small businesses and environmental investment); in Scotland, there was little likelihood of sufficient growth 'in the economy to reduce unemployment significantly in the next few years'; in the North West – *vis-à-vis* Greater Manchester – an increase in unemployment in the short term was 'the most likely prospect' (partly because of job losses in manufacturing), while in Merseyside, 'the growing disparity between the availability of jobs and the numbers seeking jobs' was of particular concern; in Wales it was clear that increased employment opportunities had to 'be created to prevent unacceptable high levels of unemployment rising further'; and in Yorkshire & Humberside, and in the West Midlands, a decrease in the rate of unemployment was dependent upon national economic recovery.

Although the report was not a forecast but a series of 'working assumptions' on which government policy could be based, the Labour Party's regional affairs spokesman, Gordon Brown, MP, declared that the report was 'a catalogue of industrial and regional decay' and that its projections were 'a grim admission that the country is divided into two separate economies' (see Naughtie, 1986). The gap between the southern regions and the rest of the country was, moreover, likely to continue well into the 1990s. Business Strategies Ltd (1989), for example, forecast that unemployment in the north would stabilize at 1.4 million by 1993 (approximately the level in 1988), but would fall to 430,000 in the south (nearly 300,000 down on 1988).

A more detailed examination of unemployment data reveals that there are further important interregional disparities. Table 2.9 shows that, both among males and females, unemployment in 1987 was generally higher in the north than in the south, but among males the greatest proportion of unemployment was long term (over three years), whereas among females, most unemployment was comparatively short term (of over one year and up to two years). The differences in the duration of unemployment are to a significant extent attributable to

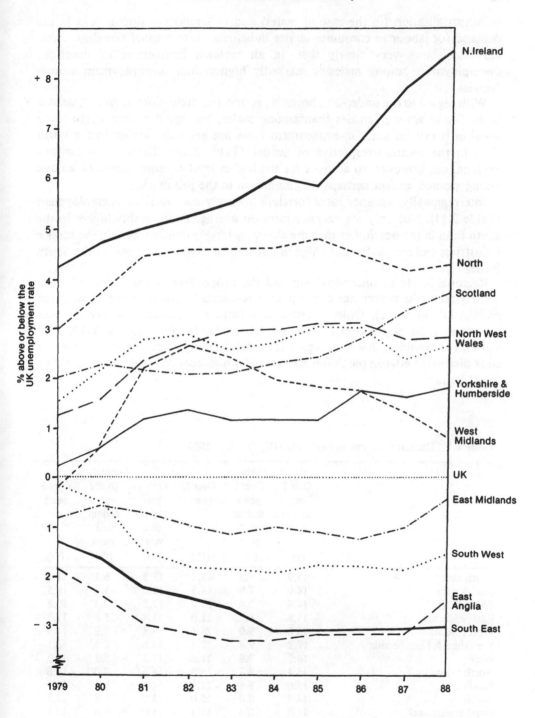

Figure 2.4 Divergence in unemployment, UK, 1979-88 (Source: Department of Employment)

deindustrialization (in the case of males) and to temporary fluctuations in the demand for labour in consumer-service industries (in respect of females). Table 2.9 also shows very clearly that, in all regions, irrespective of duration, unemployment among males is markedly higher than unemployment among females.

With regard to the under-20s, however, proportionately more unemployment is to be found among females than among males, but again a clear north-south division is evident since unemployment rates are generally higher in the north than in the south, irrespective of gender (Table 2.10). There is no obvious explanation, however, to account for the higher level of unemployment among young women, except perhaps discrimination in the job market.

Interregionally, vacancy rates correlate inversely with rates of unemployment (Table 2.11). Not only are vacancy rates on average considerably higher in the south than in the north, but they are also very largely concentrated in the service industries, and are particularly high in Greater London and the rest of the South East.

Regional levels of unemployment and the proportion of vacancies influence activity rates (the percentage of the population aged 16 and over constituting the civilian labour force). Other determinants include variations the age and sex structure of the population and the need to boost family incomes. Table 2.12 shows, for example, that with regard to female activity rates in 1987 there was a clear disparity between the South East (with its high cost of living) and the rest of Great Britain.

Table 2.9   Duration of unemployment, UK, October 1987

	Males			Females		
	Over 1 year and up to 2 years (%)	Over 2 years and up to 3 years (%)	Over 3 years (%)	Over 1 year and up to 2 years (%)	Over 2 years and up to 3 years (%)	Over 3 years (%)
South East	15.9	8.3	18.3	13.5	6.1	11.3
East Anglia	14.9	7.9	18.5	12.2	5.7	10.5
South West	14.4	7.4	17.2	12.5	5.3	10.8
East Midlands	15.8	8.7	21.6	13.4	5.8	11.7
West Midlands	15.6	9.0	28.1	14.8	7.3	14.9
Yorkshire & Humberside	17.2	9.4	22.7	14.6	6.4	12.1
Wales	14.7	7.8	21.8	13.2	5.8	11.3
North West	15.4	8.7	25.9	14.6	7.0	13.6
Scotland	15.6	8.6	21.8	14.2	6.4	11.7
North	14.3	8.4	26.0	13.9	6.8	13.7
Northern Ireland	17.0	10.4	31.3	15.0	7.0	12.7
UK	15.6	8.6	22.8	13.9	6.4	12.3

(*Source:* Central Statistical Office, *Regional Trends.*)

Table 2.10   Unemployment among the under-20s, UK, 1987, percentage

	Males			Females		
	Under 18	18–19	Total under 20	Under 18	18–19	Total under 20
South East	2.7	6.0	8.7	4.9	9.3	14.2
East Anglia	3.8	6.9	10.7	6.2	10.8	17.0
South West	3.4	6.7	10.1	5.2	9.8	15.0
East Midlands	4.0	7.3	11.3	6.4	11.4	17.8
West Midlands	4.1	7.6	11.7	7.3	13.3	20.6
Yorkshire & Humberside	5.0	8.2	13.2	8.9	13.5	22.4
Wales	4.5	7.9	12.4	7.7	13.4	21.1
North West	4.2	7.8	12.0	7.3	12.9	20.2
Scotland	5.7	7.9	13.6	9.5	12.3	21.8
North	4.2	7.4	11.6	7.6	13.8	21.4
Northern Ireland	3.5	8.4	11.9	5.2	13.8	19.0
UK	4.0	7.3	11.3	6.8	11.8	18.6

(*Source:* Central Statistical Office, *Regional Trends.*)

Table 2.11   Vacancies, Great Britain, March 1988

	Vacancies ('000)	Vacancies as a percentage of numbers unemployed	Vacancies in services as a percentage of total vacancies
South East (excl. Grt London)	59,824	23.9	75.3
(Greater London)	(31,922)	(10.1)	(80.3)
East Anglia	8,412	13.8	70.7
South West	18,496	11.9	72.9
West Midlands	22,396	8.5	66.4
East Midlands	12,441	7.7	65.1
Wales	10,617	7.5	69.4
North West	22,107	6.2	75.2
Scotland	18,514	5.8	72.3
Yorkshire & Humberside	14,666	5.8	69.7
North	10,756	5.6	72.7
Great Britain	230,151	9.3	72.8
'South'	118,654	15.1	
'North'	111,497	6.6	

(*Source:* Department of Employment.)

Interregional inequalities in production, incomes and expenditure

North-south disparities in both employment and unemployment are the principal determinants of regional inequalities in the material standard of living. Table 2.13 shows, for example, that, in the period 1983-7, only the South East and East

Table 2.12   Economic activity rates, Great Britain, 1979–87

| | Economic activity rates* | | | | Change in rate | |
| | 1979 | | 1987 | | 1979–87 | |
	Males	Females	Males	Females	Males	Females
South East	77.8	48.2	75.5	52.2	− 2.3	+ 4.0
West Midlands	79.9	49.5	75.2	49.8	− 4.7	+ 0.3
East Midlands	79.0	47.4	75.1	50.2	− 3.9	+ 2.8
North West	77.3	48.6	73.4	50.1	− 3.9	+ 1.5
Scotland	78.6	49.8	73.1	47.8	− 5.5	− 2.0
North	77.7	45.7	72.7	48.8	− 5.0	+ 3.1
Yorkshire & Humberside	77.7	46.8	72.5	49.4	− 5.2	+ 2.6
East Anglia	75.1	45.1	72.5	50.0	− 2.6	+ 4.9
South West	72.7	44.0	71.1	49.9	− 1.6	+ 5.9
Wales	75.7	41.9	68.0	43.5	− 7.7	+ 1.6
Great Britain	77.5	47.4	73.7	50.0	− 3.8	+ 2.6

Note
* The civilian labour fource as a percentage of the population aged 16 and over.

(*Source:* Department of Employment, *Labour Force Survey*.)

Table 2.13   Gross domestic product per capita, UK, 1983–7

| | Gross domestic product per capita (factor cost: current prices) | | | | | Increase |
	1983 (£)	1984 (£)	1985 (£)	1986 (£)	1987 (£)	1983–7 (%)
South East	5,028	5,367	5,910	6,575	7,183	42.9
East Anglia	4,285	4,686	5,092	5,643	6,048	41.1
East Midlands	4,199	4,463	4,911	5,378	5,762	37.2
Scotland	4,238	4,409	4,843	5,234	5,725	35.1
South West	4,152	4,467	4,868	5,380	5,699	37.3
North West	4,079	4,373	4,777	5,232	5,621	37.8
Yorkshire & Humberside	4,027	4,197	4,770	5,215	5,613	39.4
West Midlands	3,875	4,186	4,676	5,069	5,550	43.2
North	4,015	4,188	4,613	5,142	5,389	35.5
Wales	3,820	4,022	4,467	4,795	4,989	30.6
Northern Ireland	3,331	3,469	3,653	3,889	4,690	40.8
UK	4,341	4,617	5,089	5,597	6,059	39.6

(*Source:* Central Statistical Office, *Regional Trends*.)

Anglia had both a GDP per capita and an increase in the per capita GDP greater than the UK as a whole.

There was a similar disparity in the regional distribution of average gross earnings. Table 2.14 reveals that this was evident not only regionally but also at a disaggregated county level. At this more detailed scale, however, there are exceptions to the general pattern. While earnings in Grampian (Scotland), for example, are comparatively high (a situation largely attributable to high-wage employment in the oil industry), earnings are low in Cornwall, Devon and the Isle

Table 2.14   Average gross weekly earnings, UK, 1987

Area	Average gross weekly earnings (full time) 1987	
	Males on adult rates (£)	Females on adult rates (£)
Regions		
South East	254.1	167.6
Scotland	214.6	139.9
North West	212.5	138.0
South West	209.3	137.6
East Anglia	208.9	137.2
Yorkshire & Humberside	206.8	135.5
West Midlands	206.7	136.4
North	206.0	137.0
Wales	204.9	137.5
East Midlands	204.2	132.1
Northern Ireland	199.4	137.3
UK	**223.4**	**147.8**
Counties with the highest earnings		
Greater London	280.5	184.9
Berkshire	259.5	161.0
Grampian	244.8	141.8
Hertfordshire	243.9	157.3
Surrey	240.6	156.1
Buckinghamshire	235.9	161.0
Bedfordshire	232.1	144.5
West Sussex	231.9	149.9
Counties with the lowest earnings		
North Yorkshire	193.7	137.2
Isle of Wight	192.6	–
Devon	192.2	138.7
Tayside	190.6	134.1
Lincolnshire	190.2	130.2
Dumfries & Galloway	185.4	–
Shropshire	182.5	128.3
Cornwall	180.4	124.5

(*Source:* Department of Employment, *New Earnings Survey.*)

of Wight (these counties containing a disproportionate amount of low-paid employment in the tourist industry, agriculture and horticulture).

There is some evidence that the regional gap in earnings has widened. In the period 1979–87, average earnings (in cash terms) in Greater London rose by 143 per cent compared to an increase, for example, of only 109 per cent in the Northern region; and in 1986–7, alone, average earnings in the capital increased by 9.1 per cent compared to 7.7 per cent in Great Britain as a whole (Department of Employment, 1987). The widening regional gap in earnings can be attributed both to the depressing impact of unemployment upon pay rises in the north and the inflating effect of shortages in the supply of skilled labour upon wage levels in the south – particularly in Greater London.

Apart from changes in the regional pattern of average earnings, there are also widening disparities in the earnings of different categories of labour. Massey (1988) reported that the earnings of non-manual workers vary interregionally far more than the earnings of manual workers -white-collared workers in the South East (due to skill shortages) receiving considerably higher pay than similar labour in the north, and that there are greater north-south differences in the earnings of the highly paid than of the low paid. Massey suggested that these disparities imply that there is a tendency to a greater degree of income inequality (for example, between the top 10 per cent and the bottom 10 per cent of earners) in the South East, creating, since 1979, an increasingly wide 'south-south divide' not matched by a correspondingly wide 'north-north divide'. Massy also suggested that the very considerable interrogational variation in top earnings has resulted in the formation of different middle-class attitudes in different parts of the UK: for example, in the north, top earners cannot afford to leave, while in the South East they dare not leave for fear of being unable to return; and that there are very marked interregional variations in the way in which poverty is experienced since, although most welfare benefits are the same throughout the UK, high earnings in the South East pull up the cost of living to the detriment of those on poverty incomes.

Clearly, far more money is circulating in the south than in the north, maintaining a high level of demand for retail and leisure services – themselves major employers and growth areas of the economy (TCPA, 1987). Major retail developments in the north, on the other hand (such as the Metrocentre in Gateshead), arguably provide no real indication of a region's standard of living while unemployment remains high and earnings depressed (Pond, 1988).

In addition to earnings, unearned incomes must be taken into account in arriving at total personal income. Table 2.15 shows that with regard to both total personal incomes per head and (allowing for direct taxation) per-capita disposable income, the north-south disparity is even more evident than in the case of gross weekly earnings.

There were also major interregional inequalities in the distribution of personal gross incomes. Table 2.16 shows that, except in Scotland, there was a disproportionately large percentage of incomes below £5,000 per annum (in 1985-6) in all northern regions, while only in the South East (including Greater London) was the percentage of incomes of £20,000 or more significantly higher than in the UK as a whole.

The interregional disparity in earnings and inequalities in both the size and distribution of personal incomes greatly influences the pattern of household expenditure throughout the UK. Table 2.17 reveals that not only is there a clear north-south division in the average expenditure of households (with the south being able to spend significantly more than the north), but whereas housing, motoring and fares, and leisure and household goods and services generally account for a higher proportion of expenditure in the south, in the north proportionately more is spent on food, clothing and footwear, alcohol and tobacco, and fuel, light and power – a reflection of a different scale of preferences and arguably a lower standard of living.

Table 2.15   Total personal income and personal disposable income, UK, 1987

	Total personal income per head 1987 (£)	Index (UK = 100) 1987	Personal disposable income per head 1987 (£)	Index (UK = 100) 1987
South East	7,212	118.0	5,560	115.8
East Anglia	6,043	98.9	4,771	99.4
South West	5,912	96.7	4,705	98.0
Scotland	5,862	95.9	4,614	96.1
East Midlands	5,714	93.5	4,501	93.8
Yorkshire & Humberside	5,653	92.5	4,508	93.9
North West	5,601	91.7	4,457	92.8
North	5,574	91.2	4,410	91.9
West Midlands	5,436	89.0	4,303	89.6
Wales	5,144	84.2	4,138	86.2
Northern Ireland	4,979	81.5	4,011	83.6
UK	6,111	100.0	4,800	100.0

(*Source:* Central Statistical Office, *Regional Trends*).

Table 2.16   Distribution of personal gross incomes, UK, 1985-6

	Percentage of incomes in each income range		
	Under £5,000	£5,000– £19,999	£20,000 and over
South East (excl. Greater London)	21.3	68.1	10.6
Greater London	19.9	70.7	9.5
South West	24.2	69.1	6.7
Scotland	21.6	71.9	6.6
East Anglia	21.7	71.9	6.5
Northern Ireland	26.2	66.8	6.3
North West	24.1	69.7	6.2
East Midlands	22.8	71.6	5.7
Yorkshire & Humberside	23.5	71.0	5.4
Wales	25.2	69.9	4.9
West Midlands	25.5	69.6	4.8
North	25.5	69.7	4.8
UK	22.6	70.0	7.2

(*Source:* Board of Inland Revenue, *Survey of Personal Incomes*.)

Labour-supply disparity

In England and Wales, the regions that in recent years have witnessed the greatest increases in employment and business registrations, suffered the least unemployment, had proportionately the most job vacancies and enjoyed the highest standards of living, are - in general - the same regions that have had the

Table 2.17 Household expenditure, UK, 1986–7 (£ per week)

	Average expenditure per household per week	Housing	Motoring and fares	Leisure goods and services	Percentage of total					
					Household goods and services	Food	Clothing and footwear	Alcohol and tobacco	Fuel, light and power	Other goods and services
South East	219.2	18.0	15.0	15.0	12.3	17.8	7.1	6.0	4.8	4.1
South West	189.5	16.8	15.3	13.7	13.3	19.0	6.5	6.1	5.4	4.0
East Anglia	188.2	15.4	15.0	17.9	12.4	18.1	6.0	5.8	5.5	4.1
Northern Ireland	178.5	12.5	15.5	11.6	10.9	21.4	8.6	6.0	9.0	4.6
North West	172.0	16.3	14.2	12.1	11.3	19.9	7.4	8.4	6.2	4.1
East Midlands	169.8	16.7	15.4	11.9	11.3	20.3	7.0	7.3	6.0	4.0
West Midlands	166.4	16.9	14.6	11.1	10.9	20.4	8.2	7.6	6.2	4.2
Wales	163.6	14.1	14.4	11.5	10.7	21.7	7.9	8.3	6.6	4.7
Scotland	161.8	13.6	14.1	12.1	11.7	20.5	8.1	9.1	6.5	4.2
Yorkshire & Humberside	157.3	15.6	14.5	12.5	12.1	20.3	6.7	8.1	6.2	3.9
North	150.2	15.1	12.8	12.0	11.2	20.8	8.6	8.8	6.7	3.9
UK	183.2	16.5	14.7	13.4	11.9	19.3	7.3	7.1	5.7	4.1

(*Source:* Department of Employment, *Family Expenditure Surveys.*)

greatest increases in population. Thus, in the period 1981–7, the population of the southern regions increased in aggregate by 2.7 per cent, whereas the north (with its comparatively low rate of economic growth) experienced a decrease in its population of 0.1 per cent (Table 2.18).

Table 2.18   Estimated population change, England and Wales, 1981–7

Area	Population		Change
	1981 ('000)	1987 ('000)	1981–7 (%)
Regions			
East Anglia	1,894.7	2,013.7	+ 6.3
South West	4,381.3	4,588.4	+ 4.7
East Midlands	3,852.7	3,942.3	+ 2.3
South East	17,010.4	17,317.6	+ 1.8
Wales	2,813.5	2,836.2	+ 0.8
West Midlands	5,186.2	5,197.7	+ 0.2
Yorkshire & Humberside	4,918.4	4,900.2	− 0.4
North	3,117.6	3,076.8	− 1.3
North West	6,459.5	6,370.1	− 1.4
England & Wales	49,634.3	50,242.9	+ 0.9
'South'	23,286.4	23,919.7	+ 2.7
'North'	26,347.9	26,323.2	− 0.1
Principal growth counties			
Buckinghamshire	571.8	621.3	+ 8.5
Cambridgeshire	591.4	642.4	+ 8.5
Dorset	598.6	648.6	+ 8.2
Principal declining counties			
West Glamorgan	371.7	363.2	− 2.3
Cleveland	570.3	554.5	− 2.8
Merseyside	1,521.9	1,156.8	− 4.4

(*Source:* Office of Population Censuses and Surveys, *OPCS Monitor*.)

A more detailed examination of population change shows that (except for the East Midlands) the fastest-growing regions were all in the south, whereas the northern regions either grew more slowly or (as in the case of the North West, the North and Yorkshire & Humberside) witnessed a decrease in population. Similarly, at county level, the principal areas of growth (for example, Buckinghamshire, Cambridgeshire and Dorset) were also in the south, while the main areas of population decline (Merseyside, Cleveland and West Glamorgan) were dispersed throughout the north.

Since both regional variations in birth and death rates and immigration and emigration at a national level are largely insignificant, disparities in rates of regional population change must be attributable almost entirely to migration between regions. Thus, Table 2.19 shows that, in aggregate, regions in the south experienced in-migration over the period 1981–7, whereas regions in the north witnessed a net outflow of population. It is notable that the regions experiencing the greatest proportionate inflow of people (East Anglia, the South West, the

Table 2.19   Estimated net migration, England and Wales, 1981–7

Area	Net migration 1981–7	
	('000)	(%)
Regions		
East Anglia	+ 105.4	+ 5.56
South West	+ 223.3	+ 5.10
East Midlands	+ 52.8	+ 1.37
South East	+ 97.1	+ 0.57
Wales	+ 13.8	+ 0.49
Yorkshire & Humberside	− 43.5	− 0.88
West Midlands	− 61.4	− 1.18
North	− 49.6	− 1.59
North West	− 124.2	− 1.92
'South'	+ 425.8	+ 1.82
'North'	− 212.1	− 0.81
Counties with the greatest in-migration		
Dorset	+ 63.7	+ 10.64
Isle of Wight	+ 12.3	+ 10.41
East Sussex	+ 56.6	+ 8.51
Cornwall & Isle of Scilly	+ 30.8	+ 7.22
West Sussex	+ 46.1	+ 6.92
Cambridgeshire	+ 38.6	+ 6.63
Devon	+ 57.1	+ 5.91
Somerset	+ 24.2	+ 5.62
Counties with the greatest out-migration		
Staffordshire	− 5.5	− 0.54
Bedfordshire	− 3.8	− 0.74
Surrey	− 17.7	− 1.74
Co. Durham	− 12.4	− 2.03
Humberside	− 17.7	− 2.06
Mid Glamorgan	− 12.4	− 2.29
West Glamorgan	− 8.7	− 2.34
Cleveland	− 29.2	− 5.12

(*Source:* Office of Population Censuses and Surveys, *OPCS Monitor*.)

East Midlands and the South East) were also the regions that had witnessed the highest rate of population increase – and all, except the East Midlands, were in the south. Conversely, those regions witnessing an outflow of people (the North West, the North, the West Midlands and Yorkshire & Humberside) were generally the same as those showing a decrease in population – and all were northern regions. At a more localized level, again the north-south divide is apparent. The fastest-growing counties in England and Wales (1981–7) were Dorset, the Isle of Wight and East Sussex – all in the south, whereas the counties with the greatest out-migration were Cleveland, West Glamorgan and Mid Glamorgan – all in the north.

Interregional migration can create diseconomies both in the region experiencing an outflow of population and in the area witnessing an inflow.

Economic activity in the north, for example, might be adversely affected by the reverse multiplier, since 'population sucked into the booming south also sucks local spending out of regions with a consequent depressing effect on local service job creation' (Gudgin and Schofield, 1987). Conversely, in the south (where in many parts of the labour-market there are no apparent shortages) an inflow of migrants from the north can displace local job applicants, such as school-leavers (*ibid.*). Since there was an average annual outflow from the north of 50,000 people throughout much of the 1970s and 1980s and the South East ceased having a net outflow of population in the mid-1970s, the region now has 250,000 more people than would have been the case had net out-migration continued. This may be one reason accounting for an unusually high level of unemployment in the South East in the 1980s.

There is a causal relationship between interregional migration of labour and interregional disparities in employment growth, and this is particularly evident at the urban level. Tyler (1987), for example, showed that the 20 fastest-growing cities in Great Britain in the period 1971-81 were (with the exception of Aberdeen) all located in the south, whereas the bottom-20 declining urban areas (apart from those on the Thames estuary) were all in the north. Employment in the fastest-growing cities increased at an average rate of 23.3 per cent (with the number of jobs in, for example, Milton Keynes increasing by as much as 89.7 per cent and in Basingstoke by 37.4 per cent), while employment in the declining areas decreased at an average rate of 17.1 per cent, with job losses in the most depressed areas (Liverpool-Birkenhead and Sunderland) decreasing by as much as 23.9 and 31.3 per cent respectively.

In addition to interregional disparities in the inflow and outflow of population, there are interregional inequalities both in the past educational attainment of the labour force and in the distribution of occupational groups. Table 2.20 reveals that, in the late 1980s, a much higher proportion of the labour force in the South East than in the rest of the UK (and particularly in the north) had degrees or their equivalent. Although this might be partly attributable to the indigenous population of the South East achieving a comparatively higher level of educational attainment, it is probable that a major reason for this intra-regional disparity is the migration of a significant proportion of polytechnic and university graduates away from their home regions to the South East in pursuit of career opportunities. At both A- and O-levels of the GCE (or their equivalent), and at the CSE (Certificate of Secondary Education) (below grade 1), the north-south distinction was not so evident (leaving aside markedly lower levels of attainment in Northern Ireland at A-level). There were, however, very clear interregional differences in the proportion of the labour force having served an apprenticeship. Except for the East and West Midlands and Wales, a significantly greater proportion of the labour force of the north had achieved its highest qualification through the apprenticeship process. Finally, it was also the north, in general, that had the highest proportion of completely unqualified labour (although, again, Northern Ireland was an exception - but in this respect compared extremely well with the rest of the UK, having the lowest proportion of unqualified labour).

Table 2.20 Educational qualifications of the labour force, UK, 1986 (percentage*)

	Degree or equivalent	Other higher education	GCE A-level or equivalent	Apprenticeship	GCE O-level or equivalent	CSE below grade 1	None
South East	12.6	5.7	17.2	5.1	17.9	5.8	29.3
East Anglia	8.6	4.9	16.8	5.6	16.6	6.0	33.4
South West	8.2	6.5	18.0	6.7	17.7	7.2	29.6
Wales	8.1	7.1	14.5	6.9	17.3	5.1	35.7
East Midlands	7.3	5.8	17.4	6.6	14.7	6.6	35.7
Yorkshire & Humberside	7.0	6.0	15.6	8.1	16.9	5.8	35.1
Scotland	7.0	5.6	20.5	11.1	14.3	0.5	37.3
West Midlands	6.9	6.1	16.7	5.8	14.7	6.0	37.6
North West	6.9	5.7	16.4	8.0	16.4	5.9	34.6
North	6.5	6.5	15.1	10.2	15.2	6.2	35.5
Northern Ireland	6.2	6.4	12.4	12.2	14.4	4.8	20.0
UK	8.9	5.9	16.9	7.1	16.5	5.6	33.3

Note
* Percentages relate to highest qualification achieved.

(Source: Department of Employment, Labour Force Survey.)

The spatial pattern of educational attainment is significantly reflected in the interregional distribution of occupational groupings. Table 2.21 shows that in the late 1980s the south in general, and the South East in particular, had the highest proportion of managerial, professional and clerical labour, whereas (except for self-employment) there tended to be a proportionately greater representation of other occupations (largely manual) in the north, indicating a 'white collar–blue collar' divide between the south and north of Britain.

Housing – the ultimate divide?

Since housing is humankind's most important material need after food, and as there are greater interregional inequalities in house prices than in any other socio-economic indicator, then arguably the spatial disparity in house prices is the major indicator of the north-south divide (see Johnston, Pattie and Allsopp, 1988). While interregional variations in employment, unemployment, incomes and labour supply have, in aggregate, strongly influenced the geographical pattern of demand and consequently the spatial pattern of house prices, regional disparities in household growth have been the most fundamental determinant of both housing demand and housing need. Table 2.22 shows that (except for Greater London) household growth was significantly greater in the south than in the north in the period 1971–86 and was likely to continue in the same way into the next century – broadly maintaining price disparities in the medium term.

Whereas analyses of data at the level of the standard region normally show a clear north-south divide in house prices (with values peaking in the South East and being lowest in one or other northern region: Bourne, 1981; Hamnett, 1983), a more spatially detailed examination of house prices has been undertaken by Champion, Green and Owen (1988), who used building-society computerized records to aggregate house prices in each of the 280 local labour-market areas (LLMAs) in Great Britain in 1985. Champion, Green and Owen revealed that the geographical pattern of house prices was very clear cut – mean prices ranging from £66,488 in Guildford to £18,689 in Consett. Values not only peaked just to the west of the London LLMA and descended in all directions, but also the £40,000 contour coincided almost exactly with the boundary of the official South East region, although there were a handful of higher-priced areas 'widely scattered around the rest of Britain' (p. 256).

Analyses at the level of local-government areas also reveal that there was a very clear north-south divide in house prices in 1988. With regard to the 61 counties in Great Britain (and taking Northern Ireland as a whole), Table 2.23 shows that whereas the top-10 highest-priced counties were exclusively in the south, the bottom-10 areas were all in the north: mean prices ranging from £91,559 in Greater London to £25,651 in Northern Ireland. In respect of the 460 lower-level areas, the top-10 boroughs and district authorities were similarly all in the south (concentrated in Greater London and Surrey), while the bottom-10 districts were dispersed around the north. At this more detailed spatial level there was a wider disparity of house prices, ranging from £113,750 in Haringey to only £20,533 in Llanelli.

Table 2.21 Occupational groupings, UK, 1987 (percentage*)

	Managerial and professional	Clerical and related	Other non-manual	Employees Craft and similar	General labourers	Other manual	Self-employed	Unemployed or economically inactive
South East	16.5	11.1	4.1	5.8	0.2	12.8	8.3	40.5
East Anglia	13.2	7.4	5.2	7.5	0.4	14.9	8.2	42.3
South West	12.7	8.3	4.8	6.5	0.2	14.3	8.9	43.3
North West	12.7	7.9	3.5	7.4	0.4	14.3	6.0	46.1
West Midlands	12.5	7.4	3.8	8.5	0.5	15.3	5.7	45.0
Scotland	12.0	7.0	3.7	7.4	0.3	15.5	4.6	47.9
Yorkshire & Humberside	11.9	7.0	3.8	8.0	0.5	15.4	6.2	47.7
East Midlands	11.8	7.5	3.7	9.9	0.5	15.2	6.7	43.5
North	10.8	7.2	3.8	8.2	0.6	14.4	5.0	47.4
Wales	10.6	5.9	3.4	6.9	0.6	13.4	5.9	51.5
Northern Ireland	9.4	8.0	3.2	6.5	0.4	12.8	7.9	50.4
UK	13.4	8.5	3.9	7.2	0.4	14.1	6.9	44.3

Note
* Percentages relate to persons aged 16 or over.

(*Source:* Department of Employment, *Labour Force Survey*.)

Table 2.22   Regional household growth, 1971–2001

	Growth 1971–86 (%)	Number of households 1986 ('000)	Forecast growth 1986–2001 (%)
East Anglia	31.2	756	21.4
South East (excl. Grt London)	23.9	3,941	17.7
South West	24.5	1,746	16.9
East Midlands	20.3	1,489	14.4
West Midlands	15.1	1,945	9.3
Scotland	n.a.	1,193	8.8
Wales	15.2	1,049	7.8
Yorkshire & Humberside	11.9	1,882	6.2
North West	6.5	2,402	4.4
North	10.7	1,175	3.4

(*Sources:* Department of the Environment, *1985 Based Estimates of Numbers of Households*, 1988; Scottish Development Department.)

In their analysis of LLMAs, Champion, Green and Owen showed that local labour-markets played a major role in determining both the overall pattern of house-price disparities and the types of places with the highest and lowest mean prices within a region. They found that there was a particularly strong inverse correlation (– 0.661) between the unemployment rate (which indicated the general performance of the LLMA) and house prices. Similarly, at the county (or provincial) level, it is clearly evident that whereas unemployment in 1988 was highest in Northern Ireland, western Scotland, Cleveland and Durham, South Yorkshire, western Wales and Cornwall, in the same year house prices peaked in south east England (Figure 2.5).

The substantial north-south disparity in values in the late 1980s was to a considerable extent the result of differential increases in house prices over a number of years. In the period 1979–88, house prices in the UK increased dramatically. Table 2.24 shows that the largest increases in house prices over the nine years from 1979 occurred overwhelmingly in the south, particularly in the Outer Metropolitan Area, followed by East Anglia Greater London, the Outer South East and the South West (increases ranging from 234 to 220 per cent). In the north, only the East Midlands showed an increase above the UK mean.

Clearly, variable rates of house-price inflation have a very marked effect on wealth accumulation, even if the value of most of the housing stock is unrealized in the short term. Table 2.25 shows that not only was the estimated total market value of the stock significantly higher in the south than in the north in both 1981 and 1987, but also that the rate of increase in the total value of housing in the south was twice that of the increase in value in the north in the period 1981–7. Since nearly two-thirds of the housing stock in the UK was owner-occupied in 1987, it can be assumed that the market value of the owner-occupied stock was at least £553 billion (two-thirds of the total value of £830 billion). It was claimed, moreover, that a significant proportion of the accumulated capital gains in the owner-occupied sector was leaking out of the housing market through the process

Table 2.23   Localized house prices and unemployment, UK, 1988

Counties and Northern Ireland

	House prices (£)	Unemployment (%)		House prices (£)	Unemployment (%)
Top 10			*Bottom 10*		
Greater London	91,559	8.3	Northumberland	29,073	13.2
Surrey	84,240	n.a.	Co. Durham	28,594	15.0
Hertfordshire	81,337	4.4	Derbyshire	28,299	10.8
Berkshire	75,127	4.1	Dyfed	27,960	15.5
Buckinghamshire	71,127	4.4	Cleveland	27,841	18.2
Essex	69,404	7.7	Nottinghamshire	27,671	11.1
East Sussex	68,441	7.5	Gwynedd	26,678	16.8
West Sussex	67,847	3.8	South Yorkshire	26,203	15.9
Kent	63,112	8.1	Clwyd	26,096	13.5
Oxfordshire	60,009	4.2	Northern Ireland	25,651	17.3
Mean prices and unemployment	73,220	5.8		27,407	14.7

Boroughs and districts

	House prices (£)		House prices (£)
Top 10		*Bottom 10*	
Haringey	113,750	Bassetlaw	24,823
Southwark	111,864	Londonderry	24,818
Richmond	109,115	Fermanagh	24,350
Enfield	106,999	Delwyn	24,169
Barnet	106,310	Barnsley	24,101
Ealing	98,468	Newark & Sherwood	23,734
Elmbridge	96,217	Armagh	23,641
Brent	95,811	Staffordshire	
Wandsworth	94,921	Moorlands	23,175
Harrow	94,978	Easington	22,708
		Llanelli	20,533
Mean prices	102,813		23,605

Notes
1. House prices are means for the year ended 31 March 1988.
2. Unemployment data are for March 1988.
3. To ensure that comparisons are made on a 'like-with-like' basis the above prices refer exclusively to three-bedroomed semi-detached houses

(*Source:* Nationwide Anglia Building Society (1988) *House Prices: Highs and Lows – A Local View*; Department of Employment, *Employment Gazette*.)

of 'premature equity release'. It was suggested that properties were being remortgaged to facilitate an increase in the general level of consumption rather than home improvement, and in addition disposable income was being boosted by the realization of an increasing amount of inherited property. According to the Bank of England (1984), a total of £7 billion leaked out of the house-purchase/ improvement market in 1982–3 and that £6 billion of this sum was the realized value of inherited property. By 1987–8 it was claimed that the total value of

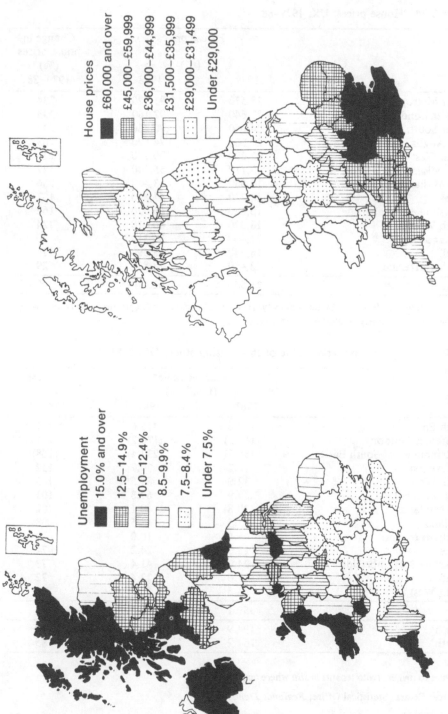

House prices

- ■ £60,000 and over
- ▦ £45,000–£59,999
- ▤ £36,000–£44,999
- £31,500–£35,999
- £29,000–£31,499
- □ Under £29,000

Unemployment

- ■ 15.0% and over
- ▦ 12.5–14.9%
- ▤ 10.0–12.4%
- 8.5–9.9%
- 7.5–8.4%
- □ Under 7.5%

Figure 2.5   Unemployment and house prices, Great Britain, 1988 (Sources: Department of Employment; Nationwide Anglia Building Society)

Table 2.24   House prices, UK, 1979–88

	House prices (£)		Change in house prices (%)
	1979	1988	1979–88
Outer Metropolitan Area	28,570	95,320	234
Greater London	27,640	92,160	233
Outer South East	24,060	80,230	233
East Anglia	20,560	74,620	263
South West	22,200	70,950	220
West Midlands	20,110	55,740	177
East Midlands	16,460	53,460	256
Wales	18,720	47,860	157
Yorkshire & Humberside	16,620	43,520	162
North West	18,270	40,910	124
Scotland	21,400	36,310	70
North	16,740	35,590	113
Northern Ireland	22,450	28,850	29
UK (average)	21,540	62,470	190

(*Sources:* Nationwide Anglia Building Society, *Housing Trends: Fourth Quarter 1979;* Nationwide Anglia Building Society, *House Prices in 1988.*)

Table 2.25   Estimated market value of the housing stock, UK, 1981–7

	Market value* (£ billion)		Increase (%)
	1981	1987	1981–7
South East	149.8	415.6	177
(Greater London)	(61.3)	(178.2)	(191)
(Remainder of South East)	(88.5)	(237.4)	(168)
East Anglia	12.2	30.7	152
South West	32.9	80.5	145
West Midlands	29.9	60.8	103
East Midlands	20.8	42.1	102
Scotland	23.5	46.5	98
Northern Ireland	5.6	10.0	79
Wales	13.9	24.2	74
Yorkshire & Humberside	24.0	41.4	73
North	13.0	22.4	72
North West	33.5	56.2	68
UK	359.1	830.3	131
'South'	194.9	526.8	170
'North'	164.2	303.5	85

Note
* Values to owners (with tenants *in situ* where appropriate).

(*Source:* Central Statistical Office, *Regional Trends.*)

was the realized value of inherited property. By 1987–8 it was claimed that the total value of residential inheritances alone had increased to £6.8 billion and it

residential inheritances alone had increased to £6.8 billion and it was predicted that by the year 2000 it would reach £8.9 billion (Hogg, 1987). Clearly, rising house prices coupled with a substantial increase in inheritance is creating 'a new class of wealth owners' (Hamnett, 1988a, p. 11) but, as Table 2.25 suggests, this new class of asset-rich households has emerged overwhelmingly in the south. If there is a continuing north-south disparity in wealth accumulation, the economic and social consequences might prove to be irreversible.

Interregional house-price variations have clearly had a substantial effect on the mobility of labour. Table 2.26 shows that if an 'average' owner-occupier had attempted to migrate in 1988 from a region of relatively high unemployment, for example, the North West, to a low-unemployment area, say Greater London, he or she would have needed to have found an additional £50,163 to have facilitated the purchase of an average-priced house in the capital, and would probably have had to increase his or her income to about £24,409 per annum (the Greater London average) to have been able to have secured this extra finance – both requirements being virtually impossible to have been met. High house prices in the South East may have thus deterred as many as 300,000 people from migrating from the north to the South East in recent years (Gudgin and Schofield, 1987), keeping unemployment in the region significantly lower than it would otherwise have been. Had house prices in the South East remained at their 1977 relative value, unemployment in the South East (according to Gudgin and Schofield) would probably have been up to 3 per cent higher and close to the national average. While house-price disparities clearly deter migration from the northern

Table 2.26 Regional house prices, incomes and unemployment, 1988

	Average house prices* (£)	Average income of borrowers* (£)	House price/ income ratio*	Unemployment† (%)
Greater London	84,853	24,409	3.48	5.8
South East (excl. Grt London)	79,953	20,180	3.96	4.5
South West	68,046	16,672	4.08	5.4
East Anglia	62,622	15,936	3.93	3.9
West Midlands	47,333	15,130	3.13	7.4
East Midlands	46,075	14,533	3.17	6.4
Yorkshire & Humberside	37,573	13,387	2.81	8.5
Wales	36,311	12,886	2.82	9.4
North West	34,690	13,257	2.62	9.7
Scotland	31,927	13,822	2.31	10.4
North	29,671	12,693	2.34	11.6
Northern Ireland	29,615	13,490	2.29	15.8
UK	52,076	16,040	3.25	9.5

Notes
* Fourth quarter.
† December.

(*Source:* Building Societies Association, *Housing Finance*; Department of Employment, *Employment Gazette*.)

regions, particularly among the manual socio-economic groups, there is conversely a net outflow of manual workers away from the high-unemployment areas of London and elsewhere in the South East because of unaffordable house prices in the region. This might benefit the South East but it undoubtedly exacerbates the unemployment problem in the more depressed parts of the rest of Britain. Managerial and professional groups in the South East, however, have normally been reluctant to migrate to the north since they could be priced out of the housing market in the South East if they wished to return. Apart from severe imperfections in the owner-occupied market, restrictions placed on council-house transfers and the shortages of private rented housing in the South East have also impeded the outflow of labour from the depressed north. While Conservative governments in the 1980s stressed the need to increase labour mobility as a means of reducing unemployment, very substantial housing constraints on migration suggest than an improvement in the economic performance of the north might provide a more effective means of bridging the north-south divide (Goddard and Coombes, 1987).

High house prices in the South East also deprive the region of an adequate inflow of managerial and professional labour from the north – perpetuating shortages of skilled manpower in the UK's most dynamic region. In the late 1980s, the South East had a gross domestic product per capita of more than 18 per cent above the national average, it contained over one-third of England's manufacturing employment and it had the highest concentration of business services, multinationals and research and development establishments (Lock, 1989). In the view of the Department of Trade and Industry, the South East (with its substantial productive capacity) is primarily in competition with the other prosperous regions of Western Europe rather than with the northern regions of the UK, which it helps to support. Industry and commerce in the South East should therefore not be prevented from achieving their full potential by an inadequate supply of skilled labour in the region. It is clearly a major cause for concern that, whereas employment forecasts suggest than an extra 77,000 dwellings per annum will be required in the South East in the 1990s (see Tyler and Rhodes, 1986), SERPLAN (an organization of all the planning authorities in the South East) propose that only 46,000 dwellings should be built each year – the difference between the two projections (31,000) possibly being the extent of annual labour vacancies in the 1990s (Lock, 1989). The shortage of affordable housing in the South East might thus become a major obstacle to the continuing economic growth of the region.

Apart from affecting relative rates of wealth accumulation, the comparatively high rate of house-price inflation in the South East might have pushed up wage levels in the region. Muellbauer (1986) suggested that since house prices were a major element in the cost of living of house-buyers, soaring house prices had put pressure on the labour-market in the mid-1980s and added 4 per cent to the level of real wages (with even greater increases accurately forecast for the late 1980s). Since wage agreements were often nationally based, he argued, house prices in the South East thus influence wage levels throughout Britain, regardless of local rates of unemployment. The view that house prices determine wage levels (rather than vice versa) could be supported by the notion that people normally buy the most expensive house they think they can afford with the largest mortgage they can raise, but then need higher wages to maintain their general standard of living. This spiral of inflation is most evident in the South East where the disequilibrium between demand and supply is clearly at its greatest. Muellbauer, Bover and Murphy (1988) further suggested that in the late 1980s households borrowing on the equity of their properties not only prolonged the consumption boom but also fuelled the rate of inflation and raised the level of import expenditure (including spending on holidays abroad), with adverse effects on the balance of payments.

Thus, as a result of very different demand influences, there were two housing markets in the owner-occupied sector in the UK in the 1980s: one in the north (greatly affected by recession) and the other in the south (associated with wealth accumulation and inflation). In different ways they both affected the 'life chances' of the households in each market. Two distinct sub-classes of owner-occupiers thereby emerged in the 1980s and, perhaps more than any other phenomenon, widened the division between north and south. The division was unlikely to have been significantly affected in 1988 by the South East experiencing its greatest net out-migration since the wartime blitz - mainly because of high house prices. Muellbauer and Murphy (1988) claimed that approximately 100,000 more people moved out of the South East in 1988 than moved into the region - and about 30,000 of this number emigrated. Since many emigrants sell their homes before leaving, it is possible that up to £5 billion was taken out of the country in 1988: equivalent to more than a third of the UK's record balance-of-payments deficit. In the 1990s, however, the southern regions are likely again to be net recipients of migrants as a result of a disproportionately fast rate of growth in employment and gross domestic product in the south, the north-south divide consequently being maintained.

While there are very evident interregional variations in housing demand (and it is these that largely influence house prices and indirectly have an impact on the macroeconomy), there are also marked disparities in housing supply. Whereas the total housing stock of the UK increased by 11 per cent in the period 1976–87, in the south the stock increased by more than 14 per cent, but in the north the rate of increase was only 9 per cent (Table 2.27). In the country as a whole, the increase in the stock was overwhelmingly attributable to house-building in the private sector - which accounted, for example, for 84 per cent of housing completions in 1987. Except for the East Midlands and Northern Ireland, the southern regions experienced the greatest proportionate rate of private-sector house-building in 1987 (notwithstanding the high cost of development land), whereas in the public sector there was an absence of any clear north-south distinction in completions.

While it is obvious that council-house sales (under the Housing Act 1980 and the Housing and Building Control Act 1984) did not increase the total stock of housing, sales quite substantially increased the size of the owner-occupied sector, although at a regionally variable rate. Dunn, Forrest and Murie (1987) revealed that of a total of 650,414 council-house sales in England (in the period 1979–85), only in the southern regions (and in the East Midlands) were sales above the

Table 2.27 Stock of dwellings and new dwellings completed, UK, 1976–87

	Stock of dwellings			New dwellings completed			
				Private sector		Public sector	
	1976	1987	Increase 1976–87	1987	Rate per 1,000 pop.	1987	Rate per 1,000 pop.
	('000)	('000)	(%)	('000)		('000)	
East Anglia	662	821	24	10.9	5.4	1.4	0.7
South West	1,614	1,872	16	19.5	4.3	2.0	0.5
East Midlands	1,395	1,576	13	17.1	4.4	2.2	0.6
South East	6,242	7,000	12	54.9	3.2	9.7	0.6
Northern Ireland	486	543	12	7.5	4.8	2.2	1.4
Wales	1,029	1,138	11	8.0	2.8	1.3	0.5
West Midlands	1,841	2,034	10	13.8	2.7	2.4	0.5
Yorkshire & Humberside	1,828	1,971	8	12.2	2.5	2.7	0.6
Scotland	1,921	2,078	8	13.8	2.7	3.8	0.8
North	1,167	1,253	7	6.7	2.2	1.7	0.6
North West	2,398	2,545	6	13.8	2.2	4.6	0.7
UK	20,613	22,830	11	178.2	3.1	34.3	0.6
'South'	8,518	9,693	14	85.3	3.6	13.1	0.5
'North'	12,099	13,137	9	92.9	2.7	21.2	0.7

(*Source*: Central Statistical Office, *Regional Trends*.)

national rate – although, in proportionate terms, they were at their lowest in Greater London (Table 2.28). Because comparatively few ex-council houses were resold shortly after acquisition, and since open-market demand for owner-occupied housing was very buoyant in the early-to-mid-1980s, it was unlikely that this increased supply of owner-occupied housing had any impact on house prices. More important was the probable effect upon housing class. If it could be shown that, by becoming owner-occupiers, former council tenants changed their class, then clearly spatial variations in council-house sales would have a spatially variable effect on housing-class formation. In general, evidence might show that there was a significantly greater degree of *embourgeoisement* of former council tenants in the south than in the north, widening still further the economic and social manifestations of the north-south divide.

Table 2.28   Council-house sales in England, 1979–85

Department of the Environment region	All sales	Sales as percentage of council stock*	Owner-occupation as percentage of total stock	
			1979	1985
South East (excl. Grt London)	140,468	16.7	63	70
Eastern	36,844	15.5	58†	66†
South West	54,305	15.0	63	69
East Midlands	63,658	14.6	59	66
West Midlands	78,379	12.5	57	63
North West (incl. Cumbria)	84,234	11.1	59‡	65‡
North (excl. Cumbria)	46,291	11.0	47§	55§
Yorkshire & Humberside	59,290	9.7	56	62
Greater London	86,954	9.6	48	55
England	650,414	12.5	57	64

Notes
* Stock is taken as stock of dwellings in 1985 plus sales.
† East Anglia standard region.
‡ North West standard region.
§ North standard region.

(*Source:* Dunn, Forrest and Murie, 1987; Department of the Environment.)

Spatial variations in both demand and supply clearly had an impact on the interregional pattern of housing tenure and changes in tenurial distribution within regions. Table 2.29 shows that, whereas owner-occupation had increased from 54 to 64 per cent of the total UK housing stock by 1987, the tenure increased to as much as 72 per cent in the South East, but to as little as 43 per cent in Scotland over the same years. Public rented housing, in contrast, declined from 32 to 26 per cent of the total housing stock in the period 1976–87, but decreased regionally at the upper level to 48 per cent in Scotland and at the lower level to 17 per cent in the South West. Outside of Greater London, private rented housing was a very minor tenure and there was an absence of any discernible north-south disparity in its distribution. In Greater London, however, private rented housing

Table 2.29 Tenure of dwellings, UK, 1976–87 (percentages)

	Owner-occupied			Private rented*			Public rented†		
	1976	1981	1987	1976	1981	1987	1976	1981	1987
South East (excl. Grt London)	61	65	72	14	12	9	24	23	18
(Greater London)	(47)	(50)	(57)	(33)	(19)	(15)	(30)	(21)	(28)
South West	62	65	71	16	14	12	22	21	17
East Midlands	57	61	69	14	11	9	29	28	22
Wales	59	63	68	13	11	9	27	27	22
East Anglia	56	60	68	17	14	12	27	25	20
North West	58	61	67	13	10	8	29	29	25
West Midlands	55	59	65	11	10	8	33	32	27
Yorkshire & Humberside	54	57	64	14	11	9	32	32	28
Northern Ireland	51	54	62	12	8	6	37	38	32
North	46	49	57	15	12	10	40	39	33
Scotland	34	36	43	12	10	8	54	54	48
UK (mean)	54	57	64	15	12	10	32	30	26

Notes
* Rented from private landlords and housing associations.
† Rented from local authorities and new-town development corporations.
(Source: Department of the Environment.)

rapidly declined from 33 to only 15 per cent of the capital's housing stock in the period 1976–87. In very broad terms, therefore, by the late 1980s, the south had become increasingly an area of owner-occupiers, the north still contained a disproportionate number of council tenants (albeit a diminishing proportion) and many households in Greater London continued to rely on a disproportionate number of rented dwellings. These patterns of distribution not only reflected regionally variable economic forces, but also (in part) determined regional standards of living.

Unequal health

Possibly the most disturbing (although, until recently, least documented) manifestation of the north-south divide is the interregional inequality of health. Referring to the spatial distribution of mortality rates, both the Black Committee (1980) and Health Education Council (1987) reported that Britain south of a line drawn from the Severn to the Wash was, by far, the healthiest part of the nation in the 1970s and 1980s. The Health Education Council pointed out that in addition to a clear north-south gradient in death rates and standard mortality ratios (SMRs), there were distinct regional differences in the disparity in health between the poorest and richest occupational classes – the health gap being greatest in the North region where the respective SMRs for men and women in classes IV and V (the partly skilled and unskilled) were 188 and 170 per cent of the SMRs for classes I and II (the professional and intermediate classes) (Table 2.30).

In addition to mortality ratios, data on disease can be used to demonstrate marked regional inequalities in health. In studies of England and Wales, Gardner *et al.* (1983) and Gardner, Winter and Barker (1984) showed that bronchitis is more common in the North West than in other regions, heart disease is largely concentrated in the North West and Wales, the greatest incidence of stomach cancer among men is likewise in these two regions, cervical cancer is particularly evident in the poorer industrial areas of the north, and lung cancer is prevalent in all urban areas – most notably in the northern conurbations and Greater London (see Haynes, 1988).

Howe (1986), in a detailed spatial analysis of premature mortality, examined the distribution of acute myocardial infarction (coronary heart disease), lung cancer and death from all causes throughout the UK at the level of district authorities. In the case of acute myocardial infarction, the areas of greatest risk for both men and women were north and west of the Severn–Wash line: Caithness, Tweeddale and Lochaber being the principal high-risk areas for men, with SMRs of 209, 197 and 195 respectively in 1980–2 (Table 2.31) – a distribution broadly replicated by the pattern of high-risk areas for women. With regard to lung cancer, and specifically among men, there is again a concentration north and west of the Severn–Wash line, the areas of greatest risk being Glasgow City, Knowsley and Middlesbrough with SMRs of 182, 179 and 167 respectively.

Taking the distribution of death from all causes, Howe indicated that, in general, the best places to live were to the south and east of a line from Gloucester to Whitby, whereas the areas in which there was the greatest likelihood of premature death were chiefly in the industrial parts of the north and west. In the

Table 2.30 Death rates and standard mortality ratios, Great Britain, 1979–80 and 1982–3

	Death rate per 1,000			Standard mortality ratios					
				Men (aged 20–59)			Women (aged 20–59)		
				Occupational class			Occupational class		
	Men	Single women	Married women	I & II	IV & V	IV & V as a percentage of I & II	I & II	IV & V	IV & V as a percentage of I & II
	(20–64)	(20–59)	(20–59)						
Scotland	6.92	1.62	2.89	87	157	180	91	141	155
North	6.43	1.56	2.50	81	152	188	80	136	170
North West	6.37	1.69	2.52	83	146	176	86	135	157
Wales	5.86	1.43	2.34	79	144	182	79	125	158
Yorkshire & Humberside	5.83	1.48	2.32	79	134	170	78	120	154
West Midlands	5.72	1.54	2.26	75	127	169	77	113	147
East Midlands	5.28	1.40	2.14	74	122	165	73	110	151
South East	4.88	1.29	1.97	67	112	167	71	100	141
South West	4.82	1.32	1.93	69	108	156	70	96	137
East Anglia	4.37	1.14	1.79	65	93	143	69	81	117
Great Britain	5.57	1.43	2.23	74	129	174	76	116	153

Note
The standard mortality ratio for all men and for all women in Great Britain is 100. The occupational classes consist of the following: I professional; II intermediate; III skilled non-manual and skilled manual (not shown); IV partly skilled; and V unskilled.

(Source: Office of Population Censuses and Surveys, 1986.)

Table 2.31   Ranking by standardized mortality ratios (SMRs) of districts in the UK with low risk and high risk of premature death, 1980–2

District	SMR	District	SMR
A			
South Norfolk	35	Afan (W. Glam.)	169
Suffolk Coastal	42	Hamilton (Strathclyde)	170
Isle of Man	44	Omagh (N. Ireland)	171
Mid Suffolk	45	Motherwell (Strathclyde)	174
Huntingdon (Cambs.)	46	Western Isles	178
Dinefwr (Dyfed)	50	Inverclyde (Strathclyde)	179
Aylesbury Vale (Bucks.)	50	Monklands (Strathclyde)	186
Coventry (W. Midlands)	50	Dungannon (N. Ireland)	194
Wychavon (Hereford &			
Worcester	51	Lochaber (Highland)	195
Chichester (W. Sussex)	51	Tweeddale (Borders)	197
Sutton (Grt London)	51	Caithness (Highland)	209
B			
Stroud (Glos.)	45	Stoke-on-Trent (Staffs.)	161
Elmbridge (Surrey)	50	Hartlepool (Cleveland)	164
Choltern (Bucks.)	50	Southwark (Grt London)	165
Waverley (Surrey)	52	Middlesbrough (Cleveland)	167
Reigate & Bansted (Surrey)	53	Knowsley (Merseyside)	179
Woodspring (Avon)	54	Glasgow City (Strathclyde)	182
C			
Hart (Hants)	59	Clydebank (Strathclyde)	143
Rochford (Essex)	61	Western Isles	145
South Norfolk	65	Kilmarnock & Loudoun	148
Elmbridge (Surrey)	66	Caithness (Highland)	149
Surrey Heath	67	Sutherland (Highland)	149
Wansdyke (Avon)	68	Badenoch & Strathspey (Highland)	151
Blaby (Leics.)	69		
South Cambridge	70	Inverclyde (Strathclyde)	152
Mid Devon	70	Glasgow City (Strathclyde)	154
South Shropshire	70	Skye & Lochalsh (Highland)	169
Wokingham (Berks.)	71	Lochaber (Highland)	182
D			
Tweeddale (Borders)	53	Londonderry (N. Ireland)	116
East Cambridgeshire	62	Cumnock & Doon Valley	
		(Strathclyde)	136
Rutland (Leics.)	66	Omagh (N. Ireland)	137
Vale of White Horse (Oxon.)	67	Cunninghame (Strathclyde)	138
Surrey Heath	67	Middlesbrough (Cleveland)	138
Wansdyke (Avon)	68	Orkney	143
Tewksbury (Glos.)	69	Monklands (Strathclyde)	144
West Somerset	69	Glasgow City (Strathclyde)	146
Mole Valley (Surrey)	69	Inverclyde (Strathclyde)	153
North Norfolk	70	Motherwell (Strathclyde)	154
UK average	100		

Note

Low risk and high risk of premature death from the following: (*A*) acute myocardial infarction (males); (*B*) lung cancer (males); (*C*) all causes (males); and (*D*) all causes (females).

(*Source*: Howe, 1986.)

case of men, SMRs ranged from 182 in Lochaber and 169 in Skye and Lochalsh to 59 in Hart and 61 in Rochford; and in the case of women SMRs ranged from 154 in Motherwell and 153 in Inverclyde to 53 in Tweeddale and 62 in east Cambridgeshire, although these extremes were not necessarily typical of high- and low-risk areas (Table 2.31 and Figure 2.6).

Gardner, Winter and Barker proffer a combination of environmental and life-style explanations of regional disparity (for example, high rates of heart disease are found in areas of soft water, relatively high rainfall and comparatively low temperatures, and are also linked to diet and a high incidence of smoking), but conclude that, overall, the chief cause of the inequality of health is the greater poverty of the north (reflected by poor housing and poor diet) and a tendency to smoke more.

There is little doubt that poor housing is associated with ill health (a relationship acknowledged over a century ago by the Public Health Acts 1872 and 1875). A number of studies have shown that SMRs are appreciably higher among people living in rented housing than among households in owner-occupied dwellings (see, for example, Fox, Jones and Goldblatt, 1984). Because rented accommodation is the least-fit tenure for human habitation and is in the greatest need of repair (*English House Condition Survey*, 1987), and since a much higher proportion of housing in the north is rented compared to the south, it follows that there is a north-south inequality in the impact of poor housing on morbidity and mortality rates. There is also evidence that poor diet and alcohol abuse are prevalent in the north. Mintel (1988) – in an analysis of interregional standards of living – reported that in the north fat is used considerably more in cooking than in the south, and that the consumption of alcohol is substantially greater: over 10 per cent of the adult population in the north drink over 15 pints of beer (or the equivalent) per week in licensed premises, compared to only 4 per cent in London and much of the South East.

Smoking-related diseases, as Roberts and Graveling (1985) found from their analysis of health-authority, local-authority and parliamentary-constituency areas, also vary in intensity interregionally, for example, from 242 deaths/ 100,000 population within the Northern region to 155/100,000 in Oxford.

It is over simple, however, to attribute most high rates of disease to any single factor. Haynes (1988, p. 505), for example, suggested that the urban pattern of lung cancer was the result of a 'combination of several weak effects [notably smoking, genetic predisposition, vitamin-A intake, hazardous industries and pollution] ... all working in the same direction'; and Howe (1986), with regard to interregional variations in SMRs, was cautious about attributing high incidences of acute myocardial infarction, lung cancer and serious disease to any one set of factors. He implied that variations in SMRs reflect complex synergistic relations between environmental influences (for example, atmospheric pollution and possibly unknown carcinogenic hazards) and life-style factors (such as poor diet, alcohol abuse, cigarette smoking and little exercise), and concludes that in the long term, 'prevention seems likely to be achieved through environmental and social management rather than through medical intervention' (*ibid.* p. 409). In the short term, however, while there is little doubt that an improved health-care

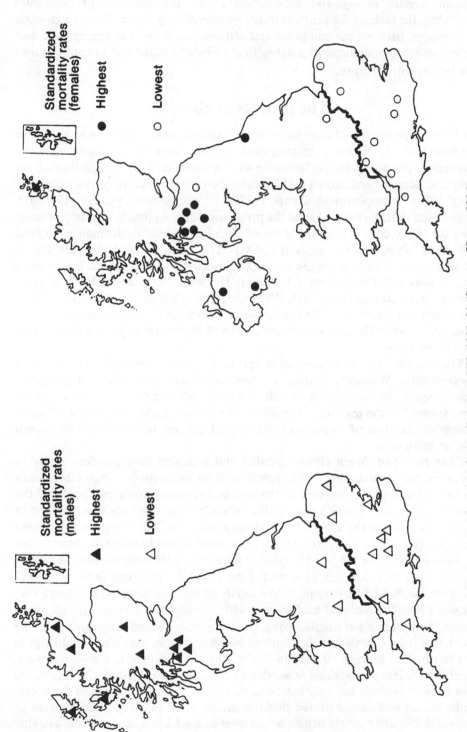

Figure 2.6 Mortality from all causes, UK, 1980–82 (Source: Howe, 1986)

system targeted at high-risk areas could reduce the incidence of premature mortality, the scale of the problem might by intensifying. Since there is a growing polarization between the employed and affluent south and the unemployed and poor north, regional disparities in health are likely to widen and to continue into the twenty-first century.

The prosperity divide

The overall prosperity of a nation or region depends very largely on its economic performance. In Britain, interregional studies show very clearly that, in aggregate, the economic performance of the south is 'superior' to that of the north. Champion and Green (1985), using percentage unemployment rates (for May 1985), unemployment change (1971–81), employment change (1978–81), population change (1971–81) and the proportion of households with two or more cars (in 1981), constructed an 'Index of Local Economic Performance' to rank 280 local labour-market areas (LLMAs). The index showed that the top-10 LLMAs were all situated in the south (and specifically in the South East region) and in rank order comprised Winchester, Horsham, Bracknell, Milton Keynes, Maidenhead, Basingstoke, High Wycombe, Aldershot & Farnborough, Bishop Stortford and Aylesbury. The top LLMA – Winchester – had the lowest level of unemployment in Great Britain (5 per cent) and 25 per cent of households owning two or more cars.

The bottom-10 LLMAs were more spread out and, in descending order, were Birkenhead & Wallasey, Irvine, Liverpool, Bathgate, Sunderland, Hartlepool, Coatbridge & Airdrie, South Shields, Mexborough and Consett – all of which were situated in the regions of the north. The bottom LLMA – Consett – had an unemployment rate of 25 per cent and only 11 per cent of its households owned two or more cars.

Champion and Green (1988) updated and extended their previous study to focus on the period since 1981, specifically to ascertain to what extent each LLMA 'shared in the recovery of the national economy from the depths of the 1979–81 recession'. Using a static index (of socio-economic indicators relating to individual years in the 1980s) and a change index (or indicators measured over four-yearly periods within the decade), an amalgamated index was constructed that showed that the 114 LLMAs lying to the south of the Severn–Wash line had a modal score of 0.525 out of a maximum possible prosperity index of 1.000, whereas the 166 LLMAs lying to the north of this line were concentrated very largely below the national median of 0.430. The top-10 LLMAs were all in the South East and East Anglia, lying in a crescent around London stretching clockwise from Crawley in the south to Newbury in the west and Cambridge in the north. The bottom 10 were in Wales, the East Midlands, the North West, Yorkshire & Humberside and in Scotland (Table 2.32 and Figure 2.7). Champion and Green's analysis not only indicated clearly the very great extent of the north-south divide and demonstrated that the divide is the primary manifestation of economic disparity across Britain at the level of the LLMA, but also showed that the gap widened in the 1980s. In the 1990s, with the lowering of European trade

Table 2.32   Best- and worst-performing local labour-markets, Great Britain, 1981–7

	Top-10 LLMAs			Bottom-10 LLMAs	
Rank	LLMA	Score	Rank	LLMA	Score
1	Milton Keynes	0.720	271	Neath	0.298
2	Newbury	0.716	272	Mansfield	0.298
3	Didcot	0.710	273	St Helens	0.298
4	Welwyn	0.709	274	Doncaster	0.293
5	Aldershot & Farnborough	0.706	275	Stranraer	0.282
6	Cambridge	0.705	276	Cardigan	0.281
7	Huntingdon	0.705	277	Pembroke	0.278
8	Hertford & Ware	0.705	278	Barnsley	0.257
9	Basingstoke	0.703	279	Mexborough	0.249
10	Crawley	0.690	280	Holyhead	0.218

(*Source*: Champion and Green, 1988.)

barriers in 1992 and the opening of the Channel tunnel later in the decade, the tilt of economic prosperity to the south will be intensified.

A further analysis of the 'prosperity divide' – by Mintel (1988) – also showed that there was a serious, even chronic, imbalance in interregional standards of living. Each of the regions of the UK were ranked in relation to a wide range of socio-economic indicators (Table 2.33) and total scores showed overall ranking. It is clearly evident that the top-three regions were in the South East, the South West and East Anglia, and the bottom-three regions were Scotland, the North and Northern Ireland. The Mintel analysis also showed that there was a very strong positive correlation between regional standards of living and the share of the Conservative vote at the 1987 general election – the political ramifications of which are examined in the next section of this chapter.

Political polarization

Although its policies in the period 1979–81 were in no small measure responsible for the worst economic recession since at least the early 1930s, the Conservatives at the 1983 general election improved upon their performance at the 1979 election by increasing their overall majority in Great Britain from 87 to 161 seats. Clearly, the Falklands war victory 'eight thousand miles from home' guaranteed the Thatcher administration 'what was not previously assured: a second term in office' (Young, 1989, p. 279). At the 1987 general election, however, the Conservatives lost ground, but were still able to secure a sizeable overall majority of 118 seats. In contrast, electoral support for the Labour Party plummeted in 1983 but recovered slightly at the 1987 election, while the Liberals and the Social Democratic Party (the Alliance during the period 1982–7) greatly improved their electoral position between 1979 and 1983. However, in 1987 they were unable to retain their earlier support. Whereas in 1983 the nationalists (Plaid Cymru and the Scottish Nationalist Party) failed to improve upon their position, at the 1987 election they made minor gains. Within this broad context, there were very marked differences in the way in which the electorate voted in the north and south (Table 2.34).

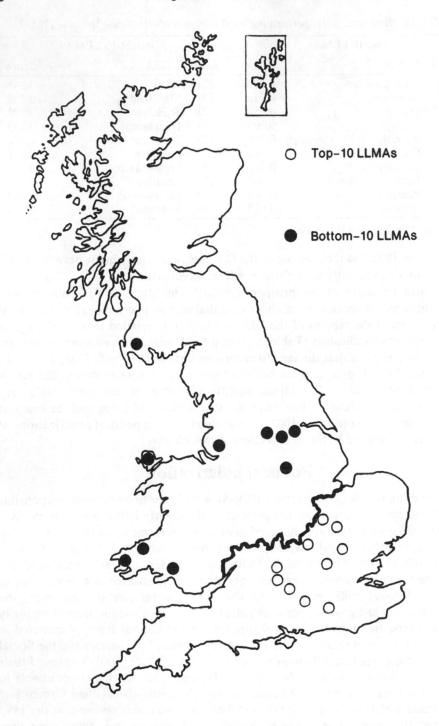

Figure 2.7 The most and least prosperous areas of Great Britain, 1981–87 (Source: Champion and Green, 1988)

Table 2.33 Interregional standard of living and electoral preference, UK, 1987

	Rank										
	SE	SW	EA	EM	NW	W	WM	Y&H	S	N	NI
Gross domestic product per capita	1	3	2	3	5	10	9	7	5	8	11
Proportion of workforce with degree or equivalent	1	2	5	6	3	7	8	10	3	11	9
Proportion of heads of households in managerial/ professional occupations	1	4	3	2	6	10	5	7	8	11	9
Average annual percentage unemployed	1	3	2	4	8	9	6	5	7	10	11
Average weekly income per household	1	3	2	9	6	8	10	11	4	7	5
Proportion of households' income derived from social-security payments	1	2	4	3	5	10	6	7	5	8	11
Average weekly household expenditure	1	3	2	9	6	8	10	11	4	7	5
Estimated gross annual income required for a moderate standard of living*	11	9	8	5	3	7†	4	1	10	1	5
Proportion of houses owner-occupied‡	1	2	5	3	6	3	7	8	10	9	—
Proportion of houses built since 1970‡	2	4	1	3	9	8	6	10	5	7	—
Average house price, third quarter 1987	11	10	9	7	5	6	8	3	4	1	2
Homes with full or partial central heating‡	1	2	5	4	10	5	8	2	11	8	5
Ownership of telephone	1	3	1	5	6	4	8	8	6	11	10
Ownership of freezer	2	1	3	3	9	5	7	7	10	6	11
Ownership of home computer	2	2	9	5	2	1	5	7	7	9	11
Cars per 1,000 population	3	1	2	5	7	6	4	8	11	10	9
Total score	41	52	65	72	98	106	107	110	111	128	119§
Conservative	2	3	1	4	6	9	5	7	10	8	—

Key
SE   South East; SW   South West; EA   East Anglia; EM   East Midlands; NW   North West;
W   Wales; WM   West Midlands; Y&H   Yorkshire & Humberside; S   Scotland; N   North;
NI   Northern Ireland.

Notes
* Gross annual income required by a family in a three–bedroomed terraced house with a 68-per-cent mortgage, 46 meals out per annum, a 1,000–cc car and no telephone.
† South Wales only.   ‡ South East excluding Greater London   § Not a full survey.
(*Source:* Mintel, 1988.)

Table 2.34   Seats won by political parties at the 1979, 1983 and 1987 general elections (Great Britain)

	Conservative			Labour			Alliance			Nationalist		
	1979	1983	1987	1979	1983	1987	1979	1983	1987	1979	1983	1987
South East	147	162	165	44	27	24	1	3	3			
East Anglia	17	18	19	2	1	1	1	1	0			
South West	42	44	44	5	1	1	1	3	3			
East Midlands	24	34	31	18	8	11	0	0	0			
West Midlands	33	36	36	25	22	22	0	0	0			
Yorkshire & Humberside	19	24	21	35	30	33	0	0	0			
North West	37	36	34	35	35	36	1	2	3			
North	6	8	8	29	26	27	1	2	1			
Wales	12	14	8	23	20	24	1	2	3	2	2	3
Scotland	23	21	10	44	41	50	3	8	9	2	2	3
Great Britain	360	397	375	260	211	229	9	21	22	4	4	6
'South'	206	224	228	51	29	26	3	7	6			
'North'	154	173	147	209	182	203	6	14	16	4	4	6

(*Source:* Johnston, Pattie and Allsopp, 1988.)

At the 1979 election, while the south was predominantly represented by the Conservatives and a sizeable proportion of constituents in the north returned Labour members, there was a notable number of Labour seats in the south. In Greater London alone, 36 of the capital's 84 seats were won by Labour candidates, and in the rest of the south Labour won 15 seats (including three in Bristol). Conversely, although Labour was strong in the north, many northern rural areas, affluent suburbs, resort and retirement centres and even some industrial towns (such as Bolton and Bury) returned Conservative members (Johnston, Pattie and Allsopp, 1988).

The election of 1983 produced a much more marked north-south division in political support, and in 1987 interregional political polarization, if anything, became even more stark. As Figure 2.8 shows, the overwhelming picture to be derived was 'one of Conservative dominance of the south and east of Great Britain ... matched by majority support for Labour in the north and west' (Johnston, Pattie and Allsopp, 1988, p.14). With regard to parliamentary seats, Table 2.34 reveals that in the South East, East Anglia and the South West, Conservative strength generally continued to increase over the whole period 1979–87, while the party made net losses in the East Midlands, Yorkshire & Humberside, the North West and Wales in 1987, and in Scotland in both 1983 and 1987. The Labour Party, reciprocally, increased its share of seats in the East Midlands, Yorkshire & Humberside, the North West, the North, Wales and Scotland in 1987, but continued to lose seats in the South East. Curiously, the relative position of the two main parties in the West Midlands did not change at the 1987 election, and in the north the Alliance and nationalist parties won only four extra seats (Table 2.34).

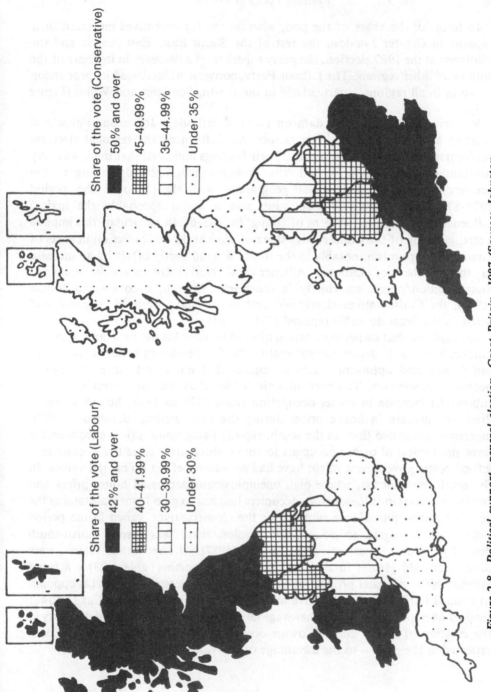

Figure 2.8   Political support, general election, Great Britain, 1987 (Source: Johnston, Pattie and Allsopp, 1988)

In terms of the share of the poll, whereas the Conservatives increased their support in Greater London, the rest of the South East, East Anglia and the Midlands at the 1987 election, the party experienced a decrease in its share of the vote in all other regions. The Labour Party, conversely, increased its proportion of votes in all regions – particularly in the North, Scotland and Wales (Figure 2.9).

Referring to opinion-poll data on political attitudes, Johnston, Pattie and Allsopp analysed the underlying factors that influenced the regional electoral performance of the political parties. Their findings indicated that there was very substantial evidence that linked 'the changing geography of voting to the country's economic and social geography [and that] ... over the period 1979–87 ... voter satisfaction/optimism was a major (probably the major) influence over changing patterns of voting' (p. 265–6). In particular, this implied a strengthening of the Conservative share of seats at the 1987 election in areas of satisfaction/optimism, notably in the south: a trend partly offset by an increase in the number of Labour, Alliance and Nationalist seats in areas of dissatisfaction/pessimism chiefly in the north. Overall, according to Mintel (1988), the Conservatives clearly did best in the 'most desirable regions' and worst in the 'least desirable regions' (Table 2.33).

Recognizing that owner-occupation offered households the best (and often the quickest) means of accumulating wealth, the Conservatives believed that voter satisfaction and optimism would be increased if more and more households became homeowners. Thatcher administrations thus not only presided over a substantial increase in owner-occupation from 1979 to 1987, but also over a dramatic increase in house prices during the same period. Johnston (1987), moreover, suggested that, in the south, rapidly rising house prices encouraged a large proportion of owner-occupiers to vote Conservative since the election of a left-of-centre government might have had an adverse effect on capital values. In the north, by contrast, where high unemployment depressed house prices and curbed increases in values, owner-occupiers had less to gain from maintaining the political status quo. Hence support for the Conservatives waned in the period 1983–7. With regard to council-house sales, there was also a north–south disparity. Dunn, Forrest and Murie (1987) showed that sales were proportionately greater in the south than in the north (Table 2.28) – a trend attributable to a greater proportion of young families, council houses (as opposed to council flats) and Conservative authorities in the south, as well as to lower rates of unemployment and higher wage levels. The resulting regional disparity in the creation of a new class of owner-occupiers might subsequently have been reflected in the polls – to the advantage of the Conservatives in the south.

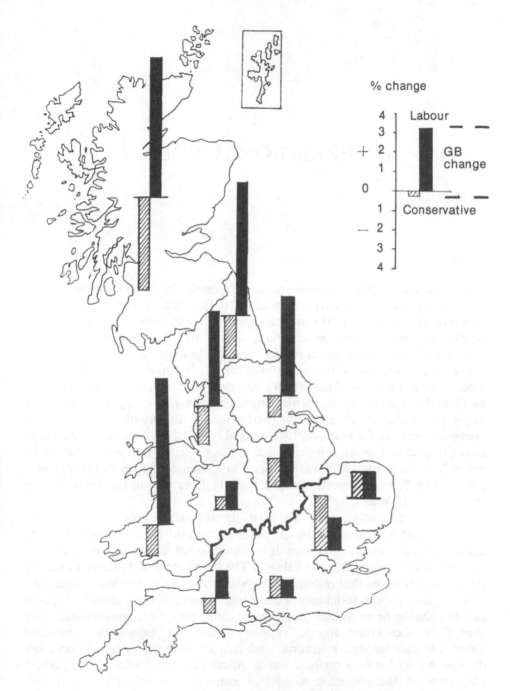

Figure 2.9 Change in the share of the vote: the general elections in Great Britain, 1983-7 (Source: Social Studies Review, September 1987)

3

INTERREGIONAL POLICY

Socio-economic indicators examined in Chapter 2 clearly showed that north-south disparities were very evident in the UK in the 1980s. There were substantial interregional variations in the growth of employment, rates of unemployment, production, incomes and expenditure, labour supply, housing, health, the standard of living and political representation. The major contrast was between the rapid development of the 'south' and the relative or absolute decline of the older industrial regions of the 'north'. Regional imbalance in the late 1980s was particularly typified by disparities in unemployment – arguably the most important indicator of economic performance. In broad regional terms, unemployment, in, for example, 1988, ranged from 5.1 per cent in East Anglia to 16.6 per cent in Northern Ireland, but although high by post-war standards, unemployment in the 1980s was relatively low when compared to the inter-war years. In 1932, unemployment ranged from 14.3 per cent in the South East to 36.5 per cent in Wales.

Since the late 1920s, policy has been directed at the problems of regional imbalance, but from time to time and particularly in the 1970s and 1980s its form has been inconsistent and its rationale confused. What, however, are the positive reasons for pursuing a policy of balance? The Department of Trade and Industry (1981) vaguely stated that regional policy was motivated by five main objectives: first, regions should ultimately achieve self-reliant growth; second, regional incomes should be at an acceptable level; third, regional unemployment rates should be acceptable; fourth, regions should not experience substantial population loss through migration; and last, unemployment rates should not diverge sharply between regions. But as Armstrong and Taylor (1985) pointed out, some of the preceding terms (for example, 'self-reliant', 'acceptable', 'substantial', etc.) mean different things to different people and would thus have a variable and possibly contradictory effect on policy formulation. The objectives of regional policy suggested in the Barlow Report (1940) are no more precise, but have had a dominant effect on regional policy over the last five decades. Regional policy, according to Barlow, was required to reduce chronic unemployment in certain heavily depressed areas and to achieve a better balance in the geographical distribution of industry to alleviate congestion and overcrowding in London (and to a lesser extent in the West Midlands).

The implications of Barlow are very clear. Higher-than-average levels of unemployment in the depressed regions of the north accompanied by low-activity rates are obvious forms of economic waste and have been partly responsible for the slow rate of national economic growth since the Second World War; at the same time, the high cost of labour, land, transportation and the provision of public services in the South East have been significant causes of inflation. Clearly the use of fiscal and monetary policies to reflate the economy has often been constrained by the risks of applying 'go' policies when the southern regions are already overheated, whereas measures to deflate the economy have been frustrated by the difficulties of applying 'stop' policies when the north is already depressed. Only with a regionally balanced economy, therefore, could macroeconomic policy ever be expected to achieve simultaneously both non-inflationary growth and full employment.

Regional policy: a consensus approach

For fifty years to the late 1970s, governments aimed to reduce the degree of regional imbalance by pursuing a policy of positive discrimination in favour of the depressed areas (see McCrone, 1969; Keeble, 1976; Maclennan and Parr, 1979; Armstrong and Taylor, 1985; Balchin and Bull, 1987). In the late 1920s, the northern regions suffered appalling levels of unemployment and poverty largely due to the depressed state of export-orientated industries (such as cotton, coal, shipbuilding and steel) concentrated in these areas. All the depressed regions had unemployment in excess of the national average of 22 per cent in 1932, but in contrast in London and the South East (where industries geared to the home market were immune from the worst effects of the world depression), unemployment was as 'low' as 13.5 and 14.3 per cent respectively.

Although the Industrial Transference Scheme of 1928 was a policy designed to 'take workers to the work' (it was intended to facilitate geographical mobility by providing grant and loan assistance to migrating labour), regional policy subsequently was largely concerned with 'taking work to the workers' through measures aimed at generating capital mobility from areas of growth (particularly the South East) to the depressed regions of the north. Commencing with the Special Areas (Development and Improvement) Act 1934, the Special Areas Reconstruction (Amendment) Act 1936 and the Special Areas (Amendment) Act 1937 (which in aggregate provided rent and rate subsidies and loans to firms in areas of high unemployment), discriminatory policy in favour of the depressed regions continued after the Second World War. Influenced by both the Barlow Report and the *White Paper on Employment Policy* (Ministry of Labour, 1944), a 'carrot-and-stick' policy of Keynesian incentives and physical constraints was introduced by the Distribution of Industry Act 1945 and the Town and Country Planning Act 1947. The 1945 Act conferred powers on the Board of Trade to provide loans and grants to firms within designated Development Areas and to facilitate the financing, building and leasing of trading estates; the 1947 Act introduced industrial development certificates (IDCs) as a means of steering industry to the development areas. IDCs were initially required for any industrial scheme in excess of 5,000 ft^2 (465 m^2) and were to be awarded discriminatively in favour of the declared areas.

The emphasis placed on capital mobility was maintained by the Distribution of Industry Act 1950, the Distribution of Industry (Industry Finance) Act 1958, the Local Employment Act 1960, the Industrial Development Act 1966 and the Industry Act 1972. Total expenditure on the major items of regional aid had increased from £22.2 million in 1962–3 to £324.2 million in 1969–70 (at 1970–1 prices), and throughout most of the 1970s – under the 1972 Act – manufacturers in the Assisted Areas were able to qualify for automatic regional development grants (RDGs) of 20-2 per cent for new industrial buildings, plant and machinery, and could write off for tax purposes 100 per cent of their capital expenditure on new plant and machinery and 44 per cent of the construction costs of new industrial buildings. Manufacturers in the Assisted Areas were also encouraged to increase or at least to maintain employment by means of Regional Employment Premiums (REPs), which were worth £3.00 per employee per week by the mid-1970s. Development controls were also being used more discriminatively. Under the 1972 Act, IDCs were no longer required in the Development and Special Development Areas, but in the Intermediate Areas they were necessary for expansion in excess of 15,000 ft^2 (1,400 m^2). In the South East they were needed for development over 5,000 ft^2 (465 m^2) and elsewhere over 10,000 ft^2 (930 m^2) – the threshold rising in 1976 to 12,500 ft^2 (1,160 m^2) and 15,000 ft^2 (1,400 m^2).

While regional policy was chiefly concerned with the location of manufacturing industry, the location of offices became the subject of legislation in the 1960s. Again, the emphasis was on the mobility of capital. In 1963 the government established the Location of Offices Bureau (LOB) to encourage firms to move out of London to take advantage of lower rents, lower labour costs, space availability and a perferable environment in decentralized locations; and under the Control of Offices and Industrial Development Act 1965, office development permits (ODPs) were introduced with the aim of steering tertiary activity to the depressed regions. At first, ODPs were required for all office development in the South East and Midlands where the proposed floor-space exceeded 3,000 ft^2 (280 m^2), but by 1970 ODPs were abolished in the West Midlands and thresholds raised to 10,000 ft^2 (930 m^2) in Greater London and the rest of the South East. However, although over 2,000 office-using firms left central and inner London between 1963 and 1979 (together with about 160,000 jobs), 75 per cent of dispersed employment was relocated in suburban areas or towns in the South East in preference to other areas in Britain, and only 9 per cent moved to the Assisted Areas (Alexander, 1979; Atkinson, 1988). Concentrated decentralization in the South East, it was claimed, made possible the retention of face-to-face contact between firms without their having to incur the high direct costs of locating in London.

To a limited extent, the government took the lead in relocating office employment. Over 55,000 civil-service jobs were dispersed to offices outside London between 1963 and 1975, and in response to recommendations in the Hardman Report (1973), a further 31,000 civil-service jobs were scheduled to move from London to the Assisted Areas. However, with cuts in public expenditure, the dispersal programme was postponed in 1979.

Almost imperceptively, regional policy throughout most of the 1970s began to break down (Townsend, 1987). In 1973, the doubling in the price of oil and the end of the 'Long Boom' heralded unemployment rising from 2.9 per cent in 1974 to 6.2 per cent in 1977, and the process of deindustrialization (underway since the mid-1960s) was associated with an 11-per-cent decline in the manufacturing labour force (1971-9). There was no longer a labour shortage in the South East and therefore the incentive for firms to relocate in areas of unemployment in the north waned. In general, investment became increasingly associated with the rationalization of production and the reduction in labour costs rather than with the direct creation of new employment (Townsend, 1987). The government not only assisted companies in the West Midlands (for example, British Leyland – now Rover), at the expense of firms and industries in the Assisted Areas, but also increasingly recognized that the inner cities as well as the regions had legitimate claims on public expenditure. While it was suggested that spatial policy should perhaps become more urban and less regional (Keeble, 1977; Fothergill and Gudgin, 1979), it was also argued that the two were not necessarily in conflict (Townsend, 1977). In retrospect, however, it was evident that regional policy was being wound down and its impact on employment creation in the 1970s was severely reduced. Marquand (1980) revealed that by the mid-1970s there had been a marked decrease in the birth rate of firms and a dramatic increase in death rates, and Killick (1983) reported that whereas there had been 1,000 openings of new manufacturing plants per annum in the late 1960s, the number had plummeted to only 300 per year by the period 1976-80. While there was a net decrease in the number of establishments in all regions, the effect upon the fragile economy of the north was particularly severe. As Keeble (1980b) had rightly argued, it became clear that regional policy 'had ceased by 1976 to exert a measurable impact on the geography of manufacturing change in Britain'. After the Conservative victory at the 1979 general election, regional policy rapidly took a new and radical direction.

The phasing-out of regional policy, 1979-87

In line with its monetarist and free-market philosophy, the Thatcher administration in June 1979 began to dismantle traditional regional policy. The belief in regional balance and in Keynesian policy as a means of eliminating spatial disparities was rejected from the outset. Planned regional aid for the period 1979-83 was immediately reduced from £842 million to £540 million (at 1980 prices), RDGs in Development Areas were cut from 20 to 15 per cent and abolished in favour of selective assistance in the Intermediate Areas (Table 3.1); the boundaries of the Assisted Areas were rolled back so that they contained only 27.5 per cent of Britain's working population as opposed to 44 per cent before (Figure 3.1); IDC controls were relaxed with the thresholds rising generally to

50,000 ft^2 (4,650 m^2); IDCs were no longer required in any of the Assisted Areas; ODPs were completely abolished. In 1982, IDCs were also abandoned altogether. 'Carrot-and-stick' policies – the basis of 'traditional regional policy' – were thus being dismantled at a time when unemployment was rising rather than falling and when manufacturing output was plummeting more rapidly than during the depression of the 1930s. Whereas unemployment in the UK as a whole increased from 4.9 to 11.7 per cent (1979–84), in the North it increased from 7.9 to 16.3 per cent, in the North West from 6.1 to 14.7 per cent and in Wales from 6.5 to 14.4 per cent.

In large part, increased unemployment was associated with record levels of redundancy. Of the new jobs provided by regional policy in the 1960s and 1970s, as many as 90,000–100,000 were lost in the period 1977–81 alone (Townsend and Peck, 1985), and during the recession of 1980 the total redundancy rate was running as high as 33,000 per month (Martin, 1982; Townsend, 1983).

The benefits of traditional regional policy

In terms of job creation over the years 1945–81, there is substantial evidence to suggest that regional policy had, on balance, been successful. Moore, Rhodes and Tyler (1981) revealed that during the twenty years after the Second World War there had been 1,057 factory 'moves' to the peripheral areas of the north –

Table 3.1   Changes in regional policy, Assisted Areas, 1972–88

	Intermediate Areas	Development Areas	Special Development Areas
Regional development grants as a percentage of factory investment:			
1972–80	20*	20	22
1980–4	Selective	15	22
1984–8	assistance only	15	Status ended
Working population as a percentage of Great Britain:			
1984	6	9	13
From 1984	20	15	Status ended
Unemployed population as a percentage of Great Britain:			
Before changes 1984†	7	10	19
After changes 1984†	25	23	Status ended
Rate of unemployment (%):			
Before changes 1984†	16.1	16.6	19
After changes†	15.9	19.6	Status ended

Notes
* Grants for machinery and plant only (buildings ineligible).
† Changes came into effect in October 1984.

(*Source*: Townsend, 1987.) Regional Policy. in W.F. Lever (ed.) *Industrial Change in the United Kingdom*, Longman, London.)

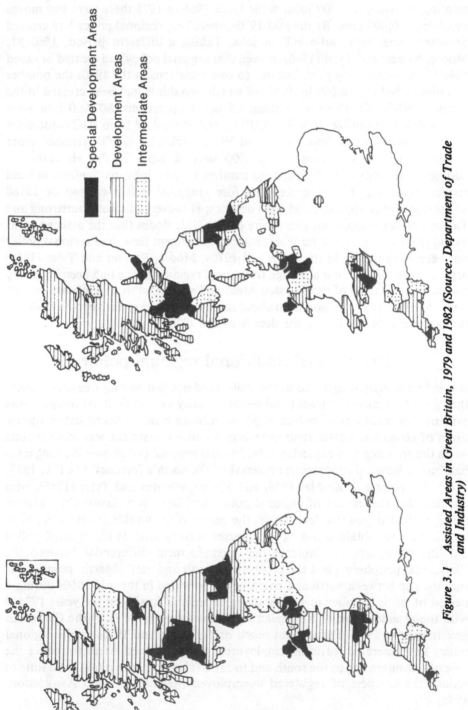

Figure 3.1 Assisted Areas in Great Britain, 1979 and 1982 (Source: Department of Trade and Industry)

Special Development Areas

Development Areas

Intermediate Areas

creating or saving 422,000 jobs, while from 1966 to 1975 there were 940 moves involving 130,000 jobs. By the mid-1970s, therefore, regional policy had created or saved more than half-a-million jobs. Taking a different period, 1960-81, Moore, Rhodes and Tyler (1986) showed that regional policy had created or saved 604,000 manufacturing jobs, but due to two recessions in the 1970s the number had diminished to 450,000 by 1981: of which two-thirds had been created in the period 1960-71. Of this total volume of new employment, 307,000 jobs were attributable to incentives (notably RDGs), 74,000 resulted from IDC constraints on job creation in the South East and West Midlands, 42,000 stemmed from regional selective assistance and 27,000 derived from REPs. However, in addition, the multiplier effect created another 180,000 jobs, particularly in local service industries. In aggregate, therefore, regional policy created or saved 630,000 jobs over approximately two decades. However, whereas Armstrong and Taylor (1987a) pointed out that 'there can be little doubt that the assisted areas would [have been] ... in a far worse state at the present time' had regional policy not been in operation in the 1960s and 1970s, Moore, Rhodes and Tyler (1986) argued more forcibly that although traditional regional policy had been both very relevant to the needs of the Assisted Areas and successful to a significant extent, it would have needed to have been about three times more effective to have solved the regional problem during the decade of the 1970s.

The costs of traditional regional policy

It was increasingly recognized in the 1980s (and not just among free-marketeers) that the achievements of traditional regional policy were limited. Although it was evident that measures to reduce regional imbalance had worked better during times of economic growth than in recession (since investment was more mobile when the economy was expanding), traditional regional policy over the long term had 'not achieved a fundamental reversal of the north's fortunes' (TCPA, 1987, p.14). In contrast to Keeble (1976) and Moore, Rhodes and Tyler (1986), who claimed that the impact of regional policy had been very favourable, Martin (1988, p. 408) argued that 'even under the generally favourable growth and policy conditions that obtained during the quarter century after 1945, regional policy did little to reduce or eliminate the unemployment differential between the "industrial periphery" and the prosperous south and east'. Martin pointed out that the gap between north and south began to widen in the mid-1960s during a period of 'strong' policy measures and widened still further in the years 1974-9 with the abandonment of Keynesian full-employment policies by the Callaghan government in 1976. Throughout much of the 1960s and 1970s, while regional policy had generated sufficient employment in the Assisted Areas to reduce the rate of out-migration to the south and to raise female activity rates, it did little to reduce the numbers of registered unemployed (Regional Studies Association, 1983).

Traditional regional policy had been targeted at manufacturing industry despite a considerable reduction in employment in this sector. In the period 1972-84, three-quarters of all regional grants went to fossil fuels, foods, metal bashing and chemicals - but all of these industries were currently experiencing large job losses. A great amount of aid was granted to companies that would have located in the north anyway, and far from creating employment, it increased the degree of capital-intensity. By mid-1984 the North Sea oil industry in Scotland had attracted very large RDGs with little direct return in jobs, and in the chemical and steel industries of Teesside the same grants yielded a net loss of jobs (Townsend, 1987): a serious cause for concern since the chemical industry had received more than 25 per cent of the £4.4 billion of RDG expenditure (1972–mid-1980s). By contrast, service industries such as finance, retailing, advertising and consultancy were generating a substantial number of jobs in London and the rest of the South East but did not qualify for regional aid, and thus had little incentive to gravitate to the north (Atkinson, 1988).

During the 1970s, regional policy largely failed to take account of the spatial implications of industrial restructuring. The north had become increasingly a branch-plant economy – an attractive location for low-wage and routine production jobs; but branch plants were remote from top management, were often strike prone, were rarely linked to local subcontractors, were neglected in terms of long-term investment and were (largely as a result) perilously exposed to economic recession (TCPA, 1987; Wray, 1987). Company mergers and takeovers stripped indigenous industries of their decision-making functions, and power became more and more centralized as company headquarters and their associated services (such as research and development) increasingly concentrated in the South East (Wray, 1987). By the mid-1980s, as many as three-quarters of the 100 top British firms had their headquarters in London, decentralization policy (through the medium of the LOB, ODPs and regional assistance) demonstrably having failed to induce them to relocate their offices beyond the South East or even outside of the capital (see Alexander, 1979; Atkinson, 1988). It was very evident, therefore, that the net effect of traditional regional policy was to concentrate decision-making, employment generation and wealth in the South East rather than to reduce the extent of regional imbalance.

A criticism often made of traditional regional policy was that it was not particularly cost-effective in creating or safeguarding jobs. Tyler (1987) showed that although the average cost of job creation/preservation over the period 1961–81 had been approximately £50,000 (at 1985 prices), this concealed a wide variation of costs relating to different forms of assistance and different industries. Employment created with the help of regional selective assistance, investment incentives (such as RDGs) and REPs cost £21,000, £31,000 and £91,000 respectively; and while (at 1982 prices) it cost an average of only £10,000 to create a job in the clothing industry, each job created in metal manufacturing cost an average of £367,000.

The impact of traditional regional policy on the South East

Whereas Moore and Rhodes (1973; 1976) and Ashcroft and Taylor (1979) suggested that it was not easy to resolve whether or not employment creation in the north had been at the expense of the South East (and London in particular),

Dennis (1980) revealed that the transfer of resources to the Assisted Areas had resulted in a loss of only 36,000 jobs over the years 1960–74 (equal to 9 per cent of the employment decline in the capital over the same period), and Fothergill and Gudgin (1982) showed that the outflow of capital from London to the Assisted Areas was no higher when regional policies were strong in the 1960s and early 1970s than when they were weak in the 1950s. After 1979, moreover, the first Thatcher government virtually gave up attempting to divert investment from the South East to the north although it encouraged overseas firms to develop production plants in the Assisted Areas – such as Nissan in Sunderland (Hall, 1981). Incentives to expand or relocate in the Assisted Areas were reduced, and ODPs and IDCs (abolished in 1979 and 1982) no longer constrained employment growth in the South East. It must be borne in mind, however, that in addition to the impact of these changes, the recession of 1979–82 (by lowering the levels of industrial and commercial investment and house-building) also limited the extent of decentralization of economic activity. In a period of economic recovery, in contrast, it could be expected that the rate of decentralization away from London and other high-cost areas of the South East would increase, regardless of the level of regional aid and the existence or absence of location controls.

The white paper of 1983

By the 1980s it was recognized that regional aid had been insensitive. It had generally not been targeted to where it could have had the greatest favourable impact (Damesick and Wood, 1987). It therefore seemed essential that regional policy should be biased towards the creation of employment rather than the promotion of capital intensity, and that simultaneously economic growth should be generated by facilitating the relocation or expansion of firms in areas that would enable them to modernize and to minimize their costs. Gudgin, Moore and Rhodes (1983) proposed that since the average cost of diverting a job to an Assisted Area amounted to as much as £40,000, a cost-per-job limit on RDGs should be introduced. Apart from stimulating investment in labour-intensive production (rather than in capital intensity), revenue released could be injected into designated growth zones in each disadvantaged region. They also pointed to the need to strengthen support for service industry in the Assisted Areas since labour demand in manufacturing was likely to remain stagnant.

Partly in response to these proposals, the white paper, *Regional Industrial Development* (Department of Trade and Industry, 1983), recommended that regional aid should be more cost-effective in creating jobs and in alleviating regional imbalances in industrial structure and employment, but it confirmed that planned regional expenditure would in total be cut by the combined effects of the revised form of grant aid and in its geographical coverage. Based largely on the white paper, reforms to regional policy were announced in November 1984. Planned spending on regional aid was reduced from £700 million in 1982–3 to £400 million by 1987–8 (at 1983 prices). As part of this reduction, RDGs were reduced from a maximum of 22 per cent to 15 per cent (Table 3.1), were awarded *either* as capital grants (up to £10,000 per job) *or* as job grants (at £3,000 per job)

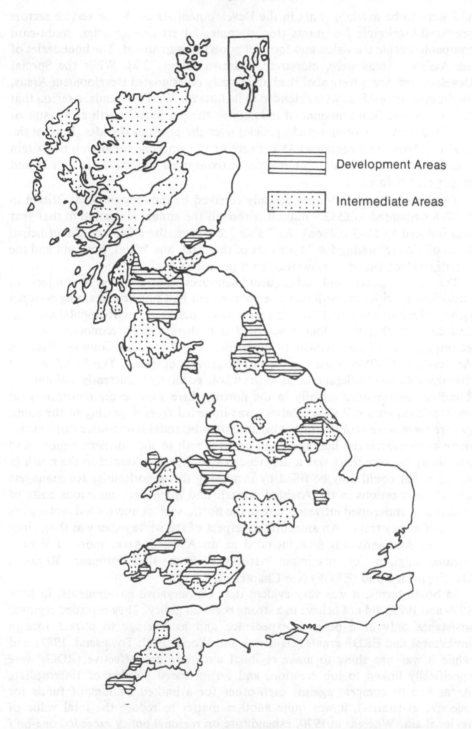

Figure 3.2   Assisted Areas in Great Britain, 1984, November (Source: Department of Trade and Industry)

and were to be available only in the Development Areas. Some service sectors previously ineligible for grants (for example, advertising agencies, credit-card companies, cable television and football pools) now qualified. The boundaries of the Assisted Areas were, moreover, redrawn (Figure 3.2). While the Special Development Areas were abolished and largely redesignated Development Areas, the Intermediate Areas were extended to include the West Midlands: a region that had lost a significant amount of industry to the peripheral north as a result of discriminatory 'carrot-and-stick' policies over the previous decades. In total the Assisted Areas now contained 35 per cent of the working population of Britain and 48 per cent of the nation's unemployed (compared respectively to only 28 and 36 per cent before).

Total preferential assistance actually received by industry in Great Britain in 1987-8 amounted to £552.9 million (although the annual allocation in that year was reduced to £283 million). As Table 3.2 shows, the RDG was the principal form of aid accounting for 70 per cent of the total, and Wales, Scotland and the North received the most assistance on a per-capita basis.

The white paper and subsequent measures, however, were subject to considerable criticism. Although the government had justified retaining regional policy (albeit in a depleted form) on essentially social and environmental grounds and declared that it no longer accepted that there was an economic case for attempting to reduce north-south disparities, the Town and Country Planning Association (1987) believed this view to be gravely mistaken. The TCPA argued that there were considerable costs to the whole economy in currently maintaining 3 million unemployed (chiefly in the north); there were severe constraints on economic expansion if skilled labour was impeded from migrating to the south (where there were skill shortages) because of substantial house-price differences; there were substantial limitations on growth both in the northern regions and nationally because there was a deficiency of enterprise potential in the north (a vacuum that could only be filled by increasing the opportunities for managers and the professions in the Assisted Areas); and there were enormous costs of supporting under-used infrastructure in the north, such as uncrowded motorways and half-empty trains. An ameliorating aspect of the white paper was that, since the West Midlands was now included as an Assisted Area, more of Britain became eligible for maximum assistance from the European Regional Development Fund (ERDF) (see Chapter 5).

In broad terms, it was very evident that Conservative governments, in both 1979 and 1983, did not believe in a strong regional policy. They regarded regional assistance only as a political expedience and as leverage to attract foreign investment and ERDF grants (Damesick and Wood, 1987; Townsend, 1987), but while it was one thing to make regional aid more cost-effective (RDGs were specifically linked to job creation, and an increased number of Intermediate Areas had to compete against each other for a limited amount of funds for selective assistance), it was quite another matter to reduce the total value of regional aid. Whereas in 1970, expenditure on regional policy exceeded one-half of 1 per cent of the gross domestic product, by 1984 it plummeted to less than one-eighth of 1 per cent of the GDP (Armstrong and Taylor, 1985).

Table 3.2 Regional preferential assistance to industry, Great Britain, 1987–8

	Regional development grants	Selective financial assistance	Expenditure by English Estates on land and buildings	Expenditure by Scottish and Welsh Development Agencies on land and factories	Development Board for Rural Wales	Highlands and Islands Development Board	Total	Per capita
	(£m)	(£m)	(£m)	(£m)	(£m)	(£m)	(£m)	(£)
Wales	76.2	28.1	–	44.2	6.5	–	132.2	47.2
North	90.7	24.4	22.6	–	–	–	108.7	35.1
Scotland	112.8	37.6	–	20.1	–	16.4	152.1	29.6
North West	68.3	19.4	13.5	–	–	–	78.6	12.3
Yorkshire & Humberside	24.2	14.1	6.7	–	–	–	38.6	7.9
West Midlands	–	18.3	0.5	–	–	–	18.8	3.6
South West	9.4	4.3	3.4	–	–	–	14.7	3.2
East Midlands	7.8	2.2	–	–	–	–	9.3	2.4
East Anglia	–	–	–	–	–	–	–	–
South East	–	–	–	–	–	–	–	–
Great Britain	389.3	148.5	46.7	64.3	6.5	16.4	552.9	10.1
'South'	9.4	4.3	3.4	–	–	–	14.7	0.6
'North'	295.3	144.2	43.3	64.3	6.5	16.4	538.2	17.1

(*Source*: Central Office of Information, *Regional Trends*.)

The measures of 1984 clearly failed to relate to the changing pattern of regional imbalance. Although recession was concentrated in the northern conurbations during the recession of 1979-81 (Owen, Coombes and Gillespie, 1983; Townsend, 1983) thereafter economic decline (and particularly deindustrialization) was disporportionately severe throughout most of the north. The scale of regional divergence was manifestly ignored by government in its redrafting of policy and, worse still, was exacerbated by Budget cuts and very poorly co-ordinated policy (Armstrong and Taylor, 1986, 1987a; Martin, 1986).

The enterprise initiative: a policy of confusion

Under the third Thatcher administration and within a programme of public-expenditure cuts, the basis of regional policy changed yet again. Whereas in 1983, the second Thatcher government had attempted to place an emphasis on cost-effectiveness, on 'value for money', the aim of the new measure set out in the white paper, *DTI - The Department for Enterprise* (Department of Trade and Industry, 1988a), was that funds should no longer be allocated to companies who could afford to pay for industrial development themselves. It was hoped, moreover, that changes would 'help spread the enterprise culture away from its natural heartland in the prosperous south east and into the regions' (Smith and Travis, 1988). RDGs, the cornerstone of aid to the Assisted Areas since 1972, were to be abolished, more money was to be available in the form of discretionary regional selective assistance, new regional-enterprise grants were to be available to small companies in the Assisted Areas, and the existing Assisted Areas map would not be changed during the lifetime of the current parliament. Although planned regional aid was to increase from £478 million in 1987-8 to £560 million in 1988-9 (at 1988 prices), it was to be cut back subsequently to £470 million in 1990-1.

Some of the 1988 measures could have been part and parcel of traditional regional policy. For example, the new regional-enterprise grants covered *either* 15 per cent (or a maximum of £15,000) of capital expenditure on new projects, *or* 50 per cent (or up to £25,000) of the cost of expansion of firms with less than 25 employees in the Development Areas; and two-thirds of the cost of business-consultancy schemes (associated with new enterprise initiatives in the Development Areas and Intermediate Areas) were also eligible for government grant. The role of the English Industrial Estate Corporation in providing industrial and commercial premises in the Assisted Areas, moreover, was to continue.

However, the major innovations of the white paper demonstrated that the Thatcher administration had virtually abdicated its balancing role in the regional arena. Although existing RDG commitments were to be honoured (RDG expenditure being maintained at £220 million in 1988-9), no further RDGs were to be awarded from 1 April 1988 - abolition being implemented by the Regional Development Grant (Termination) Act 1988. In respect of selective assistance, companies would be eligible for help only if they could satisfy the DTI that planned investment could not go ahead without government support: clearly an indication that there was to be a shift in assistance from large UK companies to small firms and overseas investors. Overall, unlike previous regional policy, the 1988 measures were overtly intended 'to help regions develop their own potential' (*The Financial Times*, 1988) - with the long-term objective of self-generated growth in the Assisted Areas.

Zero-sum gain

Whereas widening disparities in unemployment between north and south in the early-to-mid-1980s were more attributable to the recession of 1979-82 than to cuts in regional policy expenditure (Armstrong and Taylor, 1987a), thereafter changes to the grant system detailed in the 1988 white paper almost certainly had a negative effect on the volume of job creation in the Assisted Areas.

Because automatic RDGs had been abolished, civil servants and ministers now assumed direct control over the allocation of regional aid. A former Trade and Industry Secretary, Mr Leon Brittan, MP, was concerned that business people would have to go 'cap in hand' to DTI officials in the Assisted Areas for discretionary grants, while ministers themselves would have to vet all grant applications above £500,000. Mr Norman Tebbit, MP (another former Trade and Industry Secretary), also expressed doubts 'about officials becoming too deeply embroiled in taking commercial decisions' (Harrison and Brown, 1988).

Since some companies were to get assistance and others not, the new grant system was clearly incompatible with the Treasury's tenet of fiscal neutrality and it could be used excessively to attract investment from overseas with long-term cash-outflow consequences (*Guardian*, 1988a). Even before the abolition of RDGs, for every one job created in the mid-1980s there was one job lost from the number previously generated (Tyler, 1987). Under the new system of grants, the situation was almost certainly going to get worse.

Fiscal contradiction

While Conservative governments in the 1980s dramatically reduced the amount of regional aid to the north in line with their free-market and monetarist philosophy, contradictorily they had to pay out billions of pounds every year on welfare benefits to those most in need of assistance - the majority of whom were living in the north. Whereas regional-policy expenditure accounted for less than 1 per cent of planned public spending in the 1980s, social-security expenditure had increased by 30 per cent (1979-80 to 1985-6), rising to 30 per cent of total government spending by the mid-1980s. Table 3.3 not only reveals that the north received over £23 billion of cash benefits in 1986-7 (while the south benefited to the extent of only £15 billion), but also that per-capita payments were significantly higher in most northern regions than in the south. To a significant extent, therefore, both the Department of Health and Social Security and the Department of Employment (by having to pay out cash benefits) were picking up the bill left by the Department of Trade and Industry in slashing regional aid. A reduced level of regional assistance had clearly exacerbated unemployment in the north, undoubtedly increased the number of persons eligible for supplementary benefits and quite possibly increased the incidence of sickness and invalidity.

Table 3.3   Government expenditure on cash benefits, UK, 1986–7

	Total benefits	Supplementary benefits	Sickness and invalidity benefit	Unemployment benefit
£ million				
South East	10,762	2,288	552	435
North West	4,588	1,149	386	232
Scotland	3,632	787	480	210
West Midlands	3,316	836	242	143
Yorkshire & Humberside	3,294	713	283	157
South West	2,963	527	168	149
East Midlands	2,393	497	172	103
North	2,248	543	216	152
Wales	2,096	467	266	98
Northern Ireland	1,150	340	129	54
East Anglia	1,132	182	52	54
UK	37,574	8,335	2,955	1,788
'South'	14,857	2,997	772	638
'North'	22,717	5,338	2,183	1,150
£ per head				
Wales	743.2	165.7	94.2	34.8
North	729.9	176.2	70.2	49.2
Northern Ireland	728.0	217.0	82.3	34.5
North West	719.8	180.2	62.1	36.5
Scotland	709.2	153.7	93.7	41.1
Yorkshire & Humberside	672.3	145.6	57.8	32.1
South West	652.1	116.0	36.9	32.9
West Midlands	640.0	161.4	46.7	27.6
South East	623.4	132.5	32.0	25.2
East Midlands	610.4	126.9	57.8	36.2
East Anglia	568.3	91.5	43.9	27.0
UK	661.9	146.8	52.1	31.5

Note
Total benefits comprise supplementary, sickness and invalidity, and unemployment benefits; retirement, widows' and war pensions; and child benefits.

(*Source*: Central Statistical Office, *Regional Trends*.)

Notwithstanding the net flow of cash benefits to the north, Thatcher administrations in the 1980s increasingly subsidized the growing concentration of wealth in the south. Mr Michael Heseltine, MP (1988), claimed that since publicly quoted companies were given tax privileges (amounting to £6.4 billion in 1985) when taking over family firms forced on to the market-place, many family-run companies in the north were being increasingly absorbed by London-based conglomerates for tax purposes. Likewise pension funds (chiefly located in the capital) were able to exploit their tax exemptions (worth £3.5 billion in 1985) to attract savings out of wealth-creating northern firms. It was thus very evident that the tax privileges and exemptions outlined above - in total amounting to £9.9 billion in 1985 - substantially dwarfed regional assistance, which was worth only £396 million in the same year. Owner-occupiers similarly received tax privileges. Of the £4.3 billion of mortgage-interest relief received by owner-occupiers in 1985-6, a disproportionate sum went to homeowners in the South East. Exemption from both Schedule A tax (on imputed rent income) and capital-gains tax (together amounting to about £10 billion) also overwhelmingly benefited owner-occupiers in the South East, and households in the region also received a disproportionate share of discounts on council-house sales and the greatest share of improvement-grant expenditure.

The Business Expansion Scheme (BES) (introduced in 1983) likewise had a very uneven regional impact. While the BES was intended only to increase the flow of risk capital into small firms by enabling private individuals to claim tax relief at their top marginal rate on investments in new equity in unquoted companies (up to a maximum investment of £40,000 per annum), the effect of the scheme was to widen the north-south disparity in the expansion of small businesses. Whereas in the UK as a whole the amount invested under the BES increased from £105 million in 1983-4 to £157 million in 1985-6, Mason and Harrison (1989) showed that, under the scheme, both the amounts invested and the number of investee companies were disproportionately concentrated in the South East and East Anglia in the period 1983-6, and that the degree of concentration had increased over these years. Although 56 per cent of the amount invested and 50 per cent of investee companies were concentrated in the South East and East Anglia in 1983-4, the proportions increased to, respectively, 78 and 56 per cent by 1985-6. Mason and Harrison pointed out, however, that while BES investors were also disproportionately concentrated in the South East and East Anglia (these regions clearly benefiting disproportionately from tax subsidy), their degree of concentration was less than the extent to which investments and investee companies were concentrated in these regions. This suggests that not only was the BES 'contributing to a north:south flow in risk investment finance in the UK' (*ibid.* p. 52) but it was reinforcing the difficulties faced by northern firms in raising equity capital.

The March 1988 Budget further diverted wealth to the south. The abolition of rates of income tax above 40 per cent benefited high earners to the extent of £2 billion, and of this sum £1.2 billion went to people living in the South East. As Wintour (1988) revealed, although the South East contained 30 per cent of the UK's population, it received nearly 60 per cent of top rate-tax reductions. All other regions received considerably less than their share of cuts (Table 3.4). In Robson's (1988) view, the Budget had a regional impact that ran directly counter to the aims of regional policy. Calculating that the net shift of benefits from the north to the south would amount to £844 million, he suggested that the impact on the south would be a higher level of consumer spending and investment in an area already suffering from inflation.

Table 3.4   Regional distribution of top rate-tax cuts, 1988 Budget

	Total gain (£m)	Gain (%)	Population share (%)
South East	1,190	57	30
Scotland	130	6	9
South West	120	6	8
West Midlands	110	5	9
Yorkshire & Humberside	110	5	9
East Midlands	110	5	7
North West	100	5	11
East Anglia	80	4	4
North	60	3	5
Wales	60	3	5
Northern Ireland	30	1	3
'South'	1,390	67	42
'North'	710	33	58

(*Source*: Wintour, 1988.)

It was in the field of defence-equipment expenditure, however, that the economy of the south received the greatest assistance. Over the period 1979–80 to 1985–6, total defence expenditure had increased by 23 per cent and in the latter year was over £8,300 million or 14 per cent of total government spending (compared to only 11.3 per cent in 1978–9). Directly and indirectly about 625,000 people were employed in defence-equipment industries in the mid-1980s (equal to 12 per cent of total manufacturing employment) and the overall value of production (including export earnings) amounted to £10,000 million. Defence-equipment expenditure, however, was distributed very unevenly among the regions. Boddy (1987) reported that Ministry of Defence equipment expenditure (as a proportion of regional manufacturing output) ranged from as much as 18 per cent in the South East and 16 per cent in the South West to 9 per cent in both the North West and East Anglia to less than 3 per cent in Wales and Yorkshire & Humberside. Table 3.5, moreover, not only shows that the South East, by far, received the greatest share of defence-equipment expenditure in 1985–6, but also that the south as a whole received substantially more equipment spending than the north.

Employment generated from defence-equipment expenditure was clearly much greater in the south than in the north, and equipment spending in total not only offset but also completely overshadowed regional assistance – regional aid being only 7 per cent of defence-equipment expenditure in 1985–6. As Boddy pointed out, even the South West (the third-largest recipient of equipment expenditure) received almost twice as much defence-equipment spending as the amount of regional assistance in Britain as a whole. Defence-equipment expenditure thus represents a hidden regional subsidy overwhelmingly benefiting the south, but – as Boddy suggests – there is undoubtedly scope for diverting resources at the margins to satisfy the needs of both the Ministry of Defence and the peripheral regions.

It is difficult, of course, to disaggregate government spending on defence

Table 3.5   Regional preferential assistance and Ministry of Defence equipment
expenditure, Great Britain, 1985–6

	Regional preferential assistance (£m)	Defence-equipment expenditure (£m)	Defence-equipment expenditure (%)	Employment generated ('000)	Regional and defence-equipment expenditure (£m)
South East	0	4,150	50	250	4,150
North West	86	1,160	14	71	1,246
South West	12	1,000	12	61	1,012
Scotland	205	500	6	30	705
North	94	250	3	16	344
East Midlands	8	330	4	20	338
West Midlands	6	330	4	20	336
Wales	132	170	2	11	302
East Anglia	0	250	3	16	250
Yorkshire & Humberside	34	170	2	11	204
Great Britain	577	8,310	100	505	8,887
'South'	12	5,400	65	327	5,412
'North'	565	2,910	35	178	3,475

(*Sources*: Central Office of Information, *Regional Trends*; Ministry of Defence.)

equipment from public expenditure on scientific research. As well as funding
Atomic Energy and Ministry of Defence establishments such as Aldermaston,
Harwell and Winfrith, Farnborough, Malvern and Porton, the government also
finances the National Physical Laboratory in Teddington, the Meterological
Office at Bracknell, the Road Research Laboratory near Watford and numerous
agricultural research institutions. Research-council institutions, such as the Royal
Greenwich Observatory and the Rutherford Laboratory near Oxford, are also
publicly funded. It is notable that the majority of government-financed research
establishments are concentrated in a broad belt stretching across southern
England, with very few located in the north (see Davies, 1989), and it is therefore
all the more disturbing that universities in Scotland, Wales and the north of
England have incurred disproportionately large cuts in public funding in recent
years and by 1989 faced losing their research status. Clearly, with a comparatively
poor scientific and academic infrastructure, the north would find it difficult to
attract high-technology industry in the 1990s but, by comparison, in the south,
'universities, government laboratories and high tech industry will establish an
impregnable mutually supporting scientific cartel, swallowing up the lion's share
of national resources' (Davies, 1989). Only if northern universities are seen as
centres for applied research in their respective regions (and funded accordingly by
both the public and private sectors) are they likely to become catalysts for high-
technology industrial development.

Other areas of public expenditure also disproportionately benefited the South
East. British Rail (South East) and London Regional Transport, for example,
together received subsidies amounting to approximately £500 million in 1986–7 –
more than the entire regional-aid budget for that year (£424 million).

Regional planning in the South East: some major contradictions

Despite the South East benefiting disproportionately from defence procurement, spending on scientific research and public-transport subsidies, the region has suffered substantial cuts in government expenditure in recent years. From 1979 to 1983, Inner London's rate-support grant was reduced by 44 per cent in real terms, and over the decade (to 1987) district health-authority budgets were cut by 3 per cent. While investment in roads was proportionately lower than in any other region, transport subsidies were dramatically reduced following the abolition of the Greater London Council and London Transport in 1986 (Lock, 1988). Public-sector house-building was similarly a victim of cuts. For example, whereas in 1979 there were 35,000 public-sector starts in the South East, by 1984 the number had declined to 17,700 and in 1988–9 many local authorities undertook no house-building at all (Stott, 1988).

In the late 1980s, the rate of inflation increased dramatically, widening the gap between constrained current public-spending levels and more-costly regional needs. By January 1989 retail prices were rising at an annual rate of 7.5 per cent (compared to only 3.3 per cent a year earlier), and apart from its impact on the balance of payments (1988 showing a record deficit on current account of £14,270 million), inflation resulted in the government (reluctantly) having to increase public subsidies in the most overheated region: the South East (Massey, 1988). There was a need in the region to finance new infrastructure to cope with traffic congestion, and to meet the rising costs of administration as rentals increased on government properties. Escalating mortgage-interest relief also imposed a fiscal burden on the Exchequer when rates of interest rose dramatically in 1988–9.

While the South East is Britain's most prosperous region, it is 'sliding down the European prosperity league' (Stott, 1988, p. 239). In a study of the economic attributes of 102 functional urban regions (FURs) in the European Community, Cheshire, Carbonaro and Hay (1986) showed that FURs in the South East were declining at a rapid rate. London and Brighton, for example, dropped respectively from ninth and fourteenth positions in the period 1971–5 to twenty-ninth and fortieth places in the years 1981–4. It can be suggested that apart from an inadequacy of appropriate public-sector investment, a reliance on largely uncontrolled market forces and an absence of regional planning are largely attributable for this decline – a view broadly held by SEEDS (the South East Economic Development Strategy Association).

Whereas the Regional Economic Planning Council for the South East (established in 1965) assumed responsibility for planning in the region, its abolition in 1979 and effectively the simultaneous abandonment of its *Strategic Plan for the South East* left a strategic-planning vacuum (Wood, 1987). Although this was partly filled by the planning powers of the county councils and district authorities (*vis-à-vis* structure and local plans) and by letters of guidance from the Environment Secretary (concerning such matters as the protection of the Green Belt and the need to restrain the impact of the M25), throughout the 1980s there was no effective intra-regional planning framework in the South East (or elsewhere in Britain) to guide development according to broad economic and social criteria (*ibid.*).

To make matters worse, within government there was a discernible conflict of policy towards the South East. Whereas, on the one hand, the Department of the Environment limited the amount of industrial, commercial and residential development in the region and seemed committed to the maintenance of the Green Belt, on the other hand, the Department of Trade and Industry had a predilection for free-market forces, believing that if a protective barrier were to be erected around the South East, new investment (particularly from abroad) would very probably locate in mainland Europe rather than elsewhere in the UK (Lock, 1989). Clearly, a major objective of the Department of Trade and Industry was to ensure the continuing prosperity of the South East, for this (in the DTI's view) was an essential condition for the increase in the wellbeing of the economy in general. If the South East has the highest concentration of industries and companies necessary for the future vitality of the UK's economy, it is clearly essential that sufficient land is available for both industry and housing (see Department of Trade and Industry, 1988b). Under the current planning strategy for the South East, it is unlikely that such land will be released for development.

There is, of course, also an urgent need to regenerate economic activity in the north where the costs of deindustrialization are catastrophic. In short, both the south and north would benefit from a substantial increase in non-inflationary public investment and the establishment of an effective system of intra-regional planning.

Notwithstanding the very severe economic problems facing the South East, it is clear that Britain is becoming increasingly divided economically. Under Fordist conditions (where industry was mobile and regional gaps narrowed), traditional regional policy seemed appropriate. But since the mid-1970s, 'there has been no coherent spatial policy to help localities to adapt to meet the new (technology driven) circumstances' (Marquand, 1987). Fiscal contradiction and piecemeal regional assistance substantiated the view that in the late 1980s current policy was 'confused, fragmented and disorganized', and that a 'plethora of inadequately throught-out programmes [was] ... no substitute for clear co-ordination and positive policies' (North of England Regional Consortium, 1988). Thatcher administrations had clearly abandoned the fundamental 'carrot-and-stick' philosophy underlying traditional regional policy and had eschewed the whole idea of intra-regional planning – not least in the South East. Whereas free-market ideology in the 1980s resulted in considerable reductions in regional assistance (as part of a general cut-back in government expenditure), and while there was simultaneously a reluctance to use IDC controls to steer employment towards the Assisted Areas (Wray, 1987), restrictive policy in the South East not only constrained industrial and housing development, but also, through a consequential escalation in land costs, indirectly increased the rate of inflation. But, as Massey (1988) rightly pointed out, while the government had clearly 'abandoned any pretence at regional policy', it increasingly favoured targeting aid towards the inner cities.

The inner cities

For approximately thirty years after the Second World War, parts of Britain's inner cities were subject to extensive slum-clearance schemes (often involving the displacement of the local employment), massive house-building programmes and the decentralization of households to new and expanded towns and suburban estates. In the period 1945–76, 1.24 million dwellings were demolished in England and Wales, mainly in the inner cities, and from these areas about 300,000 people were rehoused under the New Towns Acts of 1946 and 1959.

In the late 1960s and in the 1970s, these policies were supplemented *first* by measures intended to improve race relations and combat urban poverty and deprivation in the inner cities. In 1968 the Urban Programme (UP) was introduced under which the government was empowered (by the Local Government (Social Need) Act 1969) to award grants of 75 per cent towards the cost of 'approved' projects bid for by local authorities and voluntary agencies (in England and Wales) in areas of 'special social need' (there were separate arrangements in Scotland). The Home Office (then deemed to be the most appropriate government department to initiate inner-city policy) subsequently established 12 Community Development Projects (CDPs) to produce detailed analyses of inner-city problems in specific geographical areas and to identify solutions that would require implementation at a central rather than a local level. Prior to their abolition in 1977, CDP teams reported that urban deprivation was not the result of social *malaise* (contrary to what was currently being claimed within the Home Office), but was the effect of unemployment, inadequate income maintenance, poor housing and a decayed environment. CDP calls for more public ownership and control of industry, and substantial changes in income maintenance and public-sector finance were largely ignored. Instead, the Comprehensive Community Programme (CCP) was initiated in 1974 to tackle urban deprivation by means of both an integrated Whitehall approach to urban problems and a new partnership between central and local government. However, since the Treasury and the spending ministries found collaboration with the Urban Deprivation Unit of the Home Office highly problematic, the CCP experiment achieved very little.

A gradual shift of emphasis from new house-building to the rehabilitation of the housing stock was the *second* modification to post-war urban-renewal policy. Under the Housing Act 1969, non-repayable discretionary grants of up to £1,200 were available for the improvement or conversion of private dwellings (and mandatory grants were similarly available for the provision of standard amenities). Equivalent sums were paid to local authorities and housing associations for the rehabilitation of public-sector stock. Under subsequent legislation (notably the Housing Acts of 1974 and 1980) the value of grants (and public-sector subsidiaries) increased substantially, although in 1974 discretionary grants became repayable if properties were sold (or left empty) within five to seven years of the receipt of a grant (a rule lifted in 1980 in the case of owner-occupied housing but retained in part in respect of landlord or developer recipients). Both in the early 1970s and again in the 1980s the number of total renovations exceeded the number of total housing starts particularly in the inner cities (see Balchin, 1989). Despite the many acknowledged short-term social and economic advantages of rehabilitation over clearance and redevelopment, critics argue that since Britain already has the oldest housing stock in Western Europe and will be unable to avoid massive expenditure on new housing in the early decades of the twenty-first century, rehabilitation is merely a vast exercise in putting off the evil day.

In the early 1970s, the Department of the Environment became increasingly involved in inner-city regeneration. Inner Area Studies (IASs) of Birmingham (Small Heath), Liverpool 8 and London (Lambeth) were commissioned by the Environment Secretary, Mr Peter Walker, in 1982. In contrast to the Home Office's socio-pathological interpretation of inner-city deprivation, the Department of the Environment acknowledged the conclusions of the IASs, which stressed that economic causes of urban decline necessitated a new policy to channel additional resources into the inner cities. The Lambeth IAS (Department of Environment, 1977a), for example, recommended that more jobs should be created or saved in inner London through industrial retention; IDC, ODP and local planning controls should be relaxed; the wholesale clearance of land (detrimental to local industry) should be abandoned; mixed land-use should be protected; and local authorities should assemble and prepare industrial land and factory premises (Buck *et al.*, 1986).

In 1976, the Department of the Environment formally replaced the Home Office as the principal government department responsible for inner-city policy. Based on the subsequent white paper, *Policy for the Inner Cities* (Department of the Environment, 1977b), the provisions of the Inner Urban Areas Act 1978 brought about an increase in public expenditure on the Urban Programme (UP) from about £30 million per annum (in 1977–8) to £165 million a year (by 1979–80). Under the Act – the *third* update in renewal policy – a total of 7 partnership authorities, 15 programme areas and 14 'other' districts were designated in England, and 5 'other' areas were designated in Wales (Table 3.6). Each of the designated authorities were given powers to provide 90-per-cent loans for the acquisition of land or site preparation, to make loans or to award grants towards the cost of setting up co-operatives or common-ownership enterprises and to declare (industrial) improvement areas (IAs). IAs were to be declared by the designated authorities to secure a stable level of employment within the older industrial and commercial areas of the inner cities. Within these areas, grants and loans were to be made available for the improvement or conversion of old industrial or commercial property. In 'special areas' within the partnerships, local authorities could make loans (interest free up to two years) for site preparation, and award grants to assist with rents. Overall assistance was to be co-ordinated by means of inner-city partnership programmes (ICPPs) or inner-area programmes (IAPs) – the former being produced by teams of civil servants and local-government officers and the latter emanating solely from programme and other designated authorities. The provisions of the 1978 Act did not apply to Scotland. Here the onus for development fell on the local and regional authorities and the Scottish Development Agency (SDA) set up in 1976, this partnership being particularly effective in planning and financing the Glasgow East Area

Table 3.6 Designated urban areas: population, 1981

Regions	Authorities	Population	Percentage of regional population
	Partnership authorities		
South East	Hackney & Islington Lambeth London Docklands: Greenwich Lewisham Newham Southwark Tower Hamlets	1,594,734	9.4
West Midlands	Birmingham	1,006,908	19.4
North West	Liverpool Manchester–Salford	510,306 692,904	18.6
North	Newcastle–Gateshead	489,332	15.7
	Programme authorities		
South East	Hammersmith	148,054	0.9
West Midlands	Wolverhampton	252,447	4.9
East Midlands	Leicester Nottingham	279,791 271,080	14.3
North West	Bolton Oldham Wirral	260,830 219,817 339,488	12.7
Yorkshire & Humberside	Bradford Hull Leeds Sheffield	457,677 268,302 704,974 536,770	40.0
North	Middlesbrough North Tyneside South Tyneside Sunderland	149,770 198,266 160,551 225,096	23.5
	Other designated authorities		
South East	Brent Ealing Haringey Wandsworth	251,257 280,042 203,175 255,723	7.3
West Midlands	Sandwell	307,389	5.9
North West	Blackburn Rochdale St Helens Sefton Wigan	141,758 207,255 189,909 300,011 308,927	17.8

Table 3.6   (*Cont.*)

Regions	Authorities	Population	Percentage of regional population
Yorkshire & Humberside	Barnsley	224,906	
	Doncaster	288,801	} 15.6
	Rotherham	251,336	
North	Hartlepool	94,359	3.0

	All designated authorities	Population	Percentage of England's population
South East		1,139,845	
West Midlands		2,573,652	
East Midlands		550,871	
North West		3,171,205	} 25.5
Yorkshire & Humberside		3,169,944	
North		1,317,374	

(*Source*: Office of Population Censuses and Surveys, *1981 Census of Population.*)

Renewal (GEAR) project in the late 1970s to 1980s. It is notable that in England and Wales, of a total of 41 designated urban areas, only 8 areas were in the south (Figure 3.3), and all of these were in inner London. Whereas it was becoming clear in the mid-1970s that the government was intending to divert public spending from the new towns to the inner cities, the House of Commons Expenditure Committee (1975) recommending that the government should review more carefully the claims new towns make on resources and that they should do more to help the urban poor, it was also very evident that regional aid was being increasingly transmitted to the north by means of inner-city policy rather than through traditional regional support. Simultaneously, within the field of interregional planning, while the Department of the Environment was expanding its role and increasing its budget, in contrast, the Department of Trade and Industry was reducing its level of regional expenditure.

Thatcherism and the inner cities

Following the return of a Conservative government in 1979, traditional regional policy increasingly gave way to a plethora of measures aimed at regenerating the inner cities and other localized problem areas. This continuing shift of emphasis may have been influenced by the view that at that time job losses in the north were mainly associated with the economic decline of the major conurbations and particularly the core areas of Clydeside, Merseyside and Tyneside, rather than attributable to the contraction of the coal, textile and shipbuilding industries – as had been the case until the mid-1960s (see Fothergill and Gudgin, 1982).

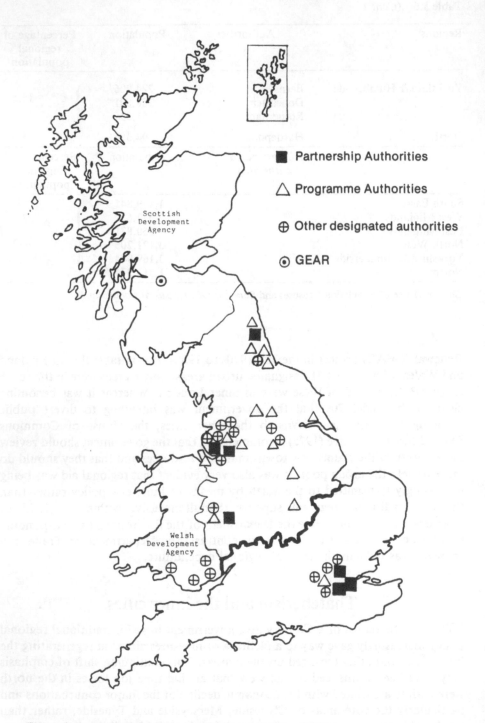

Partnership Authorities

Programme Authorities

Other designated authorities

GEAR

Scottish
Development
Agency

Welsh
Development
Agency

Figure 3.3   Designated urban areas, Great Britain, 1981

The first measure introduced by the new administration (aimed at regenerating parts of the inner city) was contained in the Local Government, Planning and Land Act 1980. Under this legislation, urban development corporations (UDCs) were designated in the London Docklands and in Merseyside to tackle problems that local authorities allegedly were unable or unwilling to remedy, and where the private sector had been deterred from investing. UDCs (usurping the planning powers of the local authorities) were intended to provide an adequate infrastructure, reclaim and service land, renovate old buildings and develop new factories – all with the aim of attracting extensive private-sector commercial and industrial development. However, although the government wished to unleash market forces, it recognized that it would need to supply a great deal of leverage finance. Whereas the London Docklands Development Corporation (LDDC) used £500 million of public funds to attract over £2 billion of private-sector investment by 1987 (a leverage ratio of 1:4), the Merseyside Development Corporation attracted only £20 million of private funds by mid-1987 in response to £140 million of public expenditure. As if to give inner-city policy a regional dimension, in 1987 further UDCs were created entirely north of the Severn–Wash line: in Trafford Park (Manchester), Sandwell (West Midlands), Teesside, Tyneside and Cardiff Bay (Figure 3.4). Each new UDC was allocated up to £160 million over the period 1987–8 to 1992–3 in the hope that high leverage ratios (of 1:4) would be attained by the early 1990s, but thereafter private finance was intended to replace Whitehall funding entirely.

UDCs, however, have generally failed to reduce local unemployment (sometimes quite the reverse), and new job vacancies have usually been filled by commuting labour. In the LDDC area, for example, whereas 2,838 'new' jobs were created between 1981 and 1987, the number of redundancies in the same period amounted to 3,344. To make matters worse, 87 per cent of the labour employed by incoming firms travelled to work from outside of the Docklands area. By 1988, unemployment in the LDDC area was consequently higher than when the LDDC was set up in 1981 (Nicholson, 1989).

In another attempt to stimulate market forces, the government declared, under the Finance Act 1980, a number of enterprise zones in which the following concessions were made: a simplified planning procedure; an exemption from rates on all commercial and industrial buildings (for 10 years from the time of a zone's declaration); a 100-per-cent first-year depreciation allowance on commercial and industrial buildings; a major reduction in government requests for statistical information; and exemption from IDC control (an irrelevant concession after 1982 due to the abolition of IDCs nationwide). By 1981, 13 enterprise zones had been declared and a further 14 in 1984, but not all were in the inner cities and, except for two, all were in the north (Figure 3.4). Enterprise zones have been subject to considerable criticism (see Armstrong and Taylor, 1985; Balchin and Bull, 1987; Balchin, Bull and Kieve, 1988). In particular, they were ineffective at generating employment and the cost of job creation was substantial. For example, Roger Tym and Partners (1984) revealed that in net terms only 5,375 jobs had been created by December 1983 at a total Exchequer cost of £252.4 million or £46,958 per job, and that of the 1,000 firms that had

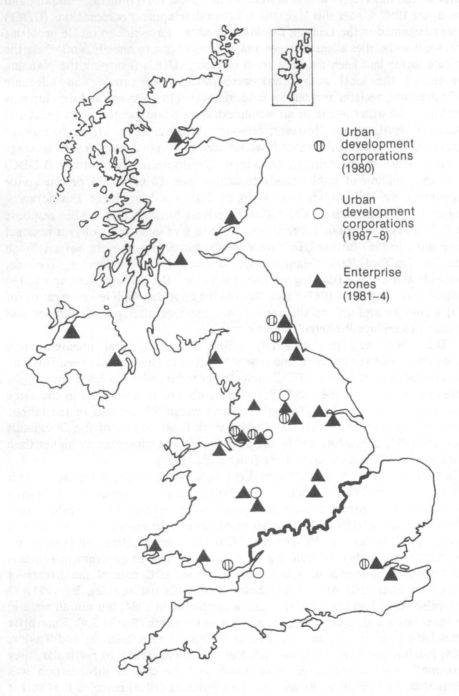

Urban
development
corporations
(1980)

Urban
development
corporations
(1987–8)

Enterprise
zones
(1981–4)

Figure 3.4   Urban development corporations and enterprise zones, UK, 1989 (Source: Department of the Environment)

been attracted to the zones by the end of 1983, three-quarters would have operated elsewhere in the same county had the zones not been designated. Even the Environment Secretary, Mr Nicholas Ridley, MP (1987) conceded that the zones had 'been rather expensive and have not given the best value for money'.

With the aim of providing an appropriate physical environment for the interplay of market forces, derelict-land grants (DLGs) (previously available only to local authorities under the Local Government Act 1972) were extended to private developers under the Local Government, Planning and Land Act 1980. To steer assistance northward, local authorities and firms were eligible for DLGs of respectively 100 per cent and 80 per cent both in the Assisted Areas and Derelict Land Clearance Areas. Elsewhere, local authorities and firms could claim grants of only 50 per cent. DLGs were particularly effective in terms of leverage. For example, in 1983-4 DLGs of £32 million were allotted to 48 public–private-sector partnership schemes that consequently produced an estimated £196 million of private development; and on the same basis an allocation of £81 million in 1987-8 will eventually attract about £400 million of private involvement.

Urban-development grants (UDGs) might possibly have been the most effective initiative to emerge from the Department of the Environment during the period of the first Thatcher government. Their purpose was to promote the economic and physical regeneration of the inner cities by levering private capital into such areas. UDGs were awarded to private firms through specified local authorities (that is, those designated under the 1978 Act and the enterprise-zone authorities). The Exchequer met 75 per cent of the grant and the local authority the remainder. The amount of the UDG was the minimum necessary to make a project commercially viable – any size or type of development being eligible. By 1986, grants totalling £120 million had levered £500 million of private investment, generating 24,000 jobs.

It became increasingly evident, however, that the private sector would be more willing to be involved in inner-city regeneration if public-sector initiatives were better co-ordinated. In 1985, therefore, five civil-service City Action Teams (CATs) were set up in the partnership areas (of Birmingham, Liverpool, Manchester, Newcastle and London) to bring together officials from the relevant government departments and managers seconded from private industry. Their purpose was to eliminate blockages in the provision of services and to ensure a more effective use of UP funds.

While urban policy was in large measure a response to economic *malaise*, the benefits of regeneration were slow to trickle down to the disadvantaged population of the inner city. In the autumn of 1980, riots broke out in the St Paul's district of Bristol, followed in 1981 by similar disturbances in Toxteth (Liverpool), Moss Side (Manchester) and Brixton (south London): all inner-city areas of deprivation. Again, in the summer of 1985, disturbances took place in Handsworth (Birmingham), Brixton and Toxteth. The government consequently attempted to design a policy that would have a more immediate and direct effect on the problems of deprivation in the inner city. Eight task forces were thus set up whereby ministers in a number of departments became responsible (under the

Environment Secretary) for improving the provision of public services in the following areas: North Peckham and Notting Hill (London), Chapeltown (Leeds), North Central Middlesborough, Highfields (Leicester), Moss Side, St Paul's and Handsworth. In each location, civil servants were allotted £1 million to support existing agencies in providing help for training, regeneration and industrial development. In 1987, a further eight task forces were set up (in Coventry, Doncaster, Hartlepool, Nottingham, Rochdale, Preston, Wolverhampton and Tower Hamlets). By the end of 1988, although task forces in total had supported 900 projects and assisted 2,000 businesses they had created or safeguarded only 4,000 jobs. While it is notable that both CATs and task forces were overwhelmingly located outside of the South East and South West (Figure 3.5), their impact on the economic problems of the north has clearly been very limited.

Towards the end of the period of the second Thatcher administration, urban-regeneration grants (URGs) were introduced under the Housing and Town Planning Act 1986, both in an attempt to increase the volume of private investment in the inner city and to by-pass local authorities. Unlike UDGs, URGs were paid directly by the Department of Environment to firms *either* to bridge the gap between the cost of development and its value on completion, *or* to provide temporary finance before any income was received from the development. While URGs were available only for substantial private-development schemes, priority was given to areas that suffered from a severe loss of employment and where there were large amounts of derelict land or disused industrial and commercial property – clearly a provision of greater benefit to the north than the south.

The third Thatcher government attempted to strengthen urban policy by giving the Department of Trade and Industry the task of spearheading the return of enterprise to the inner cities. It was believed that measures were necessary to increase the cost-effectiveness of the UP and to reduce the degree of bureaucratic constraint. But while the Department of Trade and Industry was given the responsibility of co-ordinating the work of the CATs and running the urban task forces, its inner-city budget was dwarfed by that of the Department of the Environment. By 1989–90, while the Department of Trade and Industry planned to spend only £25.8 million on its inner-city initiatives, the Department of the Environment was proposing to spend as much as £252 million on UDCs and £166 million on the UP.

In its *Action for Cities* (a programme taking over the entirety of urban aid), central government, in March 1988, announced a new UDC for Sheffield, a doubling of the area of the Merseyside Development Corporation, the introduction of mini-UDCs in Bristol, Leeds and central Manchester, a city grant to replace UDGs and URGs (payable directly to firms if leverage ratios of 1:4 could be secured) and two new CATs for Leeds and Nottingham. On the first anniversary of the launching of *Action for Cities*, three further task forces were announced: in Granby–Toxteth, Deptford (Lewisham) and Bradford. Each subsequently received £1 million in an attempt to streamline the provision of public services, but three other task forces (in Leicester, Preston and Wolverhampton) were disbanded by the end of 1989.

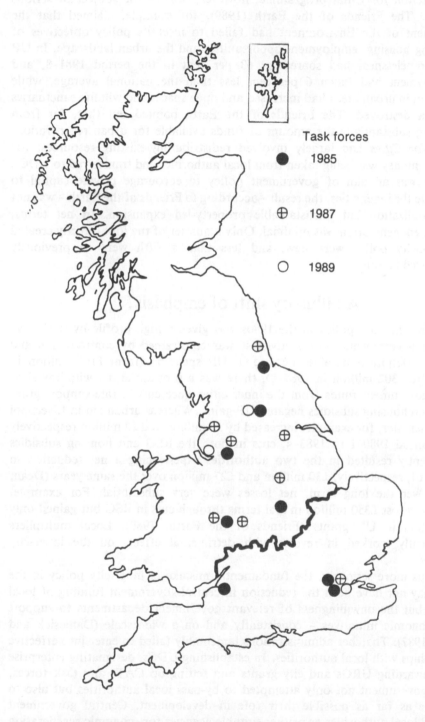

Task forces

● 1985

⊕ 1987

○ 1989

*Figure 3.5   Task forces, Great Britain, 1989 (Sources: Department of the Environment;
Department of Trade and Industry)*

The *Action for Cities* programme, however, became the subject of serious criticism. The Friends of the Earth (1989), for example, claimed that the Department of the Environment had failed to meet its policy objectives of improving housing, employment opportunities and the urban landscape. In UP areas, homelessness had soared by 63 per cent in the period 1981-8, and unemployment had fallen 6 per cent less than the national average, while dereliction in urban areas had increased and open spaces and wildlife sanctuaries had been destroyed. The Friends of the Earth pointed out that, far from increasing substantially the amount of funds available for urban regeneration, *Action for Cities* had largely involved redistributing existing resources; for example, money was being taken from local authorities and transferred to UDCs. While it was an aim of government policy to encourage private capital to regenerate the inner cities, the result – according to Friends of the Earth – was not reindustrialization but unsustainable property-led expansion. In net terms, employment generation was minimal. Only a quarter of the jobs directly created by inner-city policy were new, and less than a fifth went to previously unemployed labour.

An illusory shift of emphasis?

Although inner-city policy in the 1980s was given a high profile by successive Thatcher governments, its effectiveness was constrained by contradiction and ideology. Despite a notable increase in UP spending (from £165 million in 1979-80 to £302 million in 1987-8), there was a substantial net withdrawal of central-government funds from the inner cities since cuts in rate-support grant (RSG) and housing subsidies negated UP gains. Whereas urban aid in Liverpool and Manchester, for example, increased by £6 million and £2 million respectively in the period 1980-1 to 1983-4, cuts in both the RSG and housing subsidies consequently resulted in the two authorities experiencing a net reduction in funding of, respectively, £30 million and £27 million over the same years (Dean, 1985). Over the long term, net losses were very substantial. For example, Manchester lost £350 million in real terms through cuts in RSG but gained only £10 million in UP grants (Friends of the Earth, 1989). Local multipliers consequently worked in reverse with detrimental effects on the inner-city economy.

Perhaps more seriously, the fundamental mistake in inner-city policy in the 1980s may not have been the reduction in central-government funding of local services but the unwillingness of relevant government departments to support local economic initiatives – consistently and on a wide scale (Damesick and Wood, 1987). Thatcher administrations lamentably failed to enter into effective partnerships with local authorities. In establishing UDCs, designating enterprise zones, awarding URGs and city grants and setting up CATs and task forces, central government not only attempted to by-pass local authorities but also to eliminate as far as possible their role in development. Central government regarded local authorities as neither suitable agencies for economic regeneration (possibly due to their lack of resources and expertise – a deficiency resulting largely from cuts in central-government funding) nor as politically sympathetic. There was a widening divergence between the free-market approach of Conservative administrations and a commitment to economic planning by Labour inner-city authorities (Damesick and Wood, 1987): an ideological gap reinforced by the abolition of the Greater London Council and the metropolitan counties in 1986. But instead of replacing a so-called dependency economy (where the private sector relied upon subsidies allocated through the local authorities) with an enterprise culture (where under *laissez-faire* conditions private firms would be free to operate in an unfettered market), inner-city policy involved the creation of a new dependency – one where private capital was directly dependent upon central government leverage. Unlike local government, however, central government and its agencies were unable to respond sensitively to local needs, not least within the fields of employment and housing: a deficiency largely attributable to the demise of local accountability.

Thus, the belief that the inner cities have been substantially assisted at the expense of the regions is largely illusory. Both cities and regions have suffered severely from cuts in government expenditure and both have been disadvantageously subjected to the debilitating effects of market forces.

Whereas in 1983–4, government expenditure on regional assistance exceeded the amount of public spending on urban aid, by the end of the 1980s the inner cities were receiving significantly more assistance – in gross terms – than the regions (Table 3.7). The government for its own reasons gave inner-city policy maximum publicity and pride of place over regional policy but, as a former Trade and Industry Secretary, Mr Leon Brittan, MP (1987a), argued, the government should not have over-emphasized inner-city problems at the expense of broader causes for concern, such as the declining economy of much of the north. Although there is clearly a desperate need for the government to target substantial resources at the problems of urban deprivation, there is also an urgent need for an equivalent amount of assistance (at the very minimum) to be available to minimize the problems of regional unemployment and interregional imbalance.

Political consensus and divergence

Whereas the inter-war depression heralded the introduction of regional policy in Britain, the deep industrial recession of the early 1980s witnessed its impending collapse – although there were signs in the mid-1970s that regional assistance was on the wane. For fifty years there had been a political consensus *vis-à-vis* the need to reduce or to eliminate regional imbalance. Both major political parties were committed to a greater or lesser extent to 'one-nation' policies, and during the thirty years after the Second World War, both applied only marginally different 'carrot-and-stick' policies designed to attract resources to the north and constrain growth in the south.

It was to be expected that (when in office) Labour would adopt strong regional policies. Given that Labour was the party of the working class and the

Table 3.7 Regional and urban assistance, UK, 1983–92 (£ million)

	1983–4*	1984–5*	1985–6*	1986–7*	1987–8*	1988–9†	1989–90‡	1990–1‡	1991–2‡
Department of Trade and Industry expenditure:									
Regional assistance	426	475	396	424	283	463	455	460	410
Department of the Environment expenditure:									
Urban Development Corporations	94	88	86	89	133	234	252	270	260
Urban Programme	156	176	170	158	197	183	166	168	170
City Grant	8	16	24	24	27	34	63	70	70
Derelict-land reclamation	67	68	72	80	78	77	67	70	70
Inner-city assistance	325	348	352	351	435	528	548	578	570

Notes
(1) In 1987–8 regional development grants were phased out and replaced by regional selective assistance – the change in the form of aid temporarily reducing the total amount of regional assistance.
(2) City grants replaced urban-development grants and urban-regeneration grants in 1988.
(3) The Departments of Education and Science, Health and Transport also funded the Urban Programme – in total by approximately half the sum allocated by the Department of the Environment.
* Outturn.
† Estimated outturn.
‡ Plans.

(*Source: Treasury, The Government's Expenditure Plans, 1989–90 to 1991–92, Cmnds. 605 and 609.*)

disadvantaged, and allowing for its historical electoral base in the industrial north, the 'concept of "one nation" not only meant equality, fair shares and a classless society' (Martin, 1988, p. 407) but also by implication a significant redistribution of resources from the south to the north. Vigorous regional policies were applied during Labour periods of government. For example, under the post-war Attlee administration both the Distribution of Industry Act 1945 and Town and Country Planning Act 1947 set up policy machinery that, in modified forms, lasted until the mid-1980s; and during the Wilson governments of the 1960s, spending on regional aid increased tenfold, the Assisted Areas were extended to contain 40 per cent of the working population and regional economic-planning boards and regional economic-planning councils were set up to assume responsibility for intra-regional developing (Parsons, 1986; Martin, 1988).

A belief in 'one nation', however, was first articulated not by the Labour Party but by Benjamin Disraeli. In 1845, in *Sybil*, he wrote of Britain consisting of two nations – one rich and privileged, and one poor and underprivileged, and he saw that it was his role to improve the living standards of the industrial and urban poor and so enable the Conservative Party (which he was later to lead) to claim that it was the party of national unity. In the depression of the 1920s and 1930s, although *laissez-faire* generally prevailed in government, the Conservatives nevertheless inaugurated regional policy by establishing the Industrial Transference Board in 1928, and as the largest party in the National administration, Conservatives were largely responsible for the Special Areas legislation of the 1930s. Notable Conservatives, moreover, such as Robert Boothby, Anthony Eden and Harold Macmillan, espousing 'one-nation' philosophies, advocated a marked increase in government intervention and the adoption of economic planning (see Martin, 1988); and in particular Macmillan (alluding to the unacceptability of high unemployment in the north) urged further positive intervention to eliminate the excesses of regional imbalance (see Macmillan, 1938). From the Churchill administration of 1951-5 to the Heath government of 1970-4, Conservative regional policy was essentially Keynesian and Beveridge in form (the party having eschewed *laissez-faire* dogma), and Conservatives 'shared with Labour a commitment [to a policy geared to the maintenance of] full employment and the provision of an extensive welfare state' (Martin, 1988, p. 407). This was all the more remarkable since many Conservatives may have believed that there was no need to gain support in the north to achieve electoral success. It was probable that there were enough winnable seats in the south to assure the party victory in most elections without having to design regional policies to attract the floating voter of the north.

With the oil-price hike of 1973, the long post-war boom finally came to an end throughout much of the industrialized world. Restrictive monetary policy was widely introduced to curb inflation but it severely constrained economic growth, external trade, industrial output and employment of each country in which it was applied. In the UK, moreover, as part of a disinflationary package the Labour government in the period 1975-9 'imposed the largest post-war cut in real spending on regional aid, a reduction in real terms of 50 per cent (about £700 million) from 0.7 to 0.3 per cent of GDP' (Martin, 1988, p. 408). In contrast to its period in office in the 1960s, Labour in the 1970s clearly failed to make the elimination of the north-south divide a priority. But while it may have been broadly true that the concept of 'one nation' was sacrificed on the altar of price stability, it might also have seemed to a section of the party that the working classes of the north had been betrayed in an attempt to generate (or to retain) political support in the south. Discounting the possibility of disaffection in the north, the Wilson–Callaghan administrations might well have believed that support in the few Labour strongholds in the south would dissipate (with severe electoral effects) if, during a period of disinflation, the north failed to take its full share of cuts - for example, in the field of regional aid.

Concentrated industrial action over pay policy during the 1978-9 'winter of discontent' was a precursor to the return of a Conservative government in May 1979. Within months, the first Thatcher administration removed exchange controls and exposed the UK economy (and particularly the fragile north) to the full impact of external trends upon growth, incomes and employment (Gudgin and Schofield, 1987). The outflow of capital abroad, moreover, had damaging effects on investment in the UK, not least where it was most needed: the areas of industrial recession in the north. In the years that followed, Thatcher governments attempted to rid the economy of inflation and to stimulate 'the regenerative powers of efficient private capitalism by controlling the money supply and fostering free competition' (Martin, 1988, p. 410).

In 1979–81, a policy of high interest rates (aimed at 'squeezing inflation out of the system') resulted in sterling becoming over-valued by at least 40 per cent, with severe effects on exports, output, company solvency and employment. Keynesian policy was subsequently turned on its head and instead of reflating the economy, the Chancellor, Sir Geoffrey Howe, introduced Draconian cuts in public expenditure (Gudgin and Schofield, 1987). The UK, meanwhile, suffered the deepest industrial recession this century. For example, manufacturing output plummeted by as much as 15 per cent in 1980 compared to a decrease of only 6.9 per cent during the worst year of the inter-war depression. According to the Organization for Economic Co-operation and Development, while the macroeconomic policy of the Thatcher government did not cause the recession (or even less credibly bring about the deindustrialization of the north or London), it doubled its severity (Greater London Council, 1985a). Traditional forms of regional assistance, moreover, were being superseded both by 'central-government localism' (with an emphasis on UDCs, enterprise zones and the extension of the UP), and a reliance on the small-business sector, 'aided by tax reductions and a plethora of schemes to promote enterprise' (Martin, 1988, p. 411).

Supporters of *laissez-faire*, both inside and outside of government, argued that a freeing of the market would eliminate north-south disparities. It was suggested, for example, that the abolition of national wage agreements and their replacement by regional or local wage settlements would not only result in unemployed labour in the north being able to price itself into work but also, if more and more firms subsequently moved northwards to benefit from lower wage costs, regional house-price variations would be narrowed (Martin, 1988). This scenario assumes, of course, that trade unions would be willing to enter into regional wage bargaining (even if it could be shown that, at some reasonable level of wage, unemployment would be reduced to negligible proportions – an unlikely situation), and that firms (including multinationals) and financial institutions would be prepared to forgo the perceived benefits of a southern location.

Although the recession of the early 1980s was followed by gradual reflation, slow recovery and increasing exports (helped by both a falling pound and rising productivity), by 1987 manufacturing output was barely back to the level it had reached in 1979, while the diminution of regional policy and a greater dependence on the market almost certainly widened the north-south divide.

In one respect it could be argued that the Thatcher administrations of the 1980s had taken the passive view that little could be done about regional imbalance. During the first half of the decade, the government may have assumed that, at best, regional policy could not create jobs but only divert them, but in the late 1980s it seemed that even this half-hearted approach had largely given way to *laissez-faire* (Breheny, Hall and Hart, 1987). An alternative interpretation of Conservative policy was that there was a deliberate attempt to promulgate an enterprise economy, a 'nouveau bourgeoise "yuppie" society dominated by a new class of owner-occupiers and shareholders – concentrated overwhelmingly in the South East' (Massey, 1988). However, while the professional and managerial classes (social economic groups (SEGs) 1–4) constituted a disproportionately large percentage of the population of the South East, this, according to Massey, was not the result of market forces or of southern hard work but directly attributable to public policy and class power.

During the 1980s, financial services in the City were prioritized over manufacturing in the north. Employment in banking, finance and insurance in the South East increased markedly between 1979 and 1987 (see Table 2.5), whereas high interest rates and an over-valued pound squeezed the regions – manufacturing employment plummeting in the north (see Table 2.1). While disinflationary cuts in government spending had a devastating effect on the economy of the north, a high level of public expenditure on, for example, defence equipment, British Rail (South East) and the construction of the M11, the M25 and London's third airport all generated economic activity in the South East and particularly benefited SEGs 1–4.

In 1981, SEGs 1–4 accounted for up to 34 per cent of households in the Outer Metropolitan Area of the South East, whereas in the north (except in Yorkshire & Humberside) SEGs 1–4 constituted no more than 22.9 per cent of households. A significant proportion of professional and managerial households in the South East, however, were not indigenous but had migrated from elsewhere in response to labour shortages and increasingly concentrated job opportunities in the region. The South East, consequently, was the only region where both male and female earnings were above the UK average in 1987 (see Table 2.14). Clearly, SEGs 1–4 in the South East exercised a considerable influence over Conservative policy. While it was by no means certain that the majority of the adult population within this segment of the electorate would necessarily vote Conservative, by applying measures that positively discriminated in favour of the South East in general, and

SEGs 1-4 in particular, Thatcher administrations did their best to ensure that this would happen both at the 1983 and 1987 general elections.

Whereas the Labour Party, when in office in the 1970s, had partly turned its back on regional policy, during the 1980s it evolved a new approach to the regions based on decentralization. It saw the need for regional government to bring together the many strands of decision-making vested in central-government departments and other public bodies, and to eliminate inefficiency resulting from the lack of co-ordination in regional resource expenditure. It recognized that decision-making must be made accountable at regional level, but above all Labour recognized that the prevailing free-market approach to regional imbalance was doomed to failure. Along with other policy options, a more detailed consideration of Labour's regional policy is included in Chapter 5.

4

THE NORTH–SOUTH DIVIDE – MYTH?

For more than fifty years, regional policy in Great Britain has been largely a response to perceived economic and social disparities between north and south. Regional data on which this perception is based can, however, be seriously misinterpreted. First, the distinction between booming and decaying areas is often masked, if socio-economic indicators are considered on a regional scale rather than examined at a county or urban level; second, intra-regional disparities might be as great (or greater) than interregional variations; third, inner-city decay is apparent in most of the regions of Britain (but this is not evident from an examination of regional statistics alone); fourth, since data might not clearly indicate nationwide or cross-regional processes, interpretation might be grossly over-simplified or at worst incorrect; and finally, whereas the South East is the country's most prosperous region, it contains (in Greater London) the largest single concentration of deprivation in Western Europe.

Clearly, there is no neat geographical line separating the depressed urban areas from booming towns and cities. Instead of a north-south divide, it could be suggested that the essential contrast is primarily between the older industrial areas and the new centres of growth, wherever they may be situated. For example, industrial Cornwall and the Medway towns of Kent are 'old', while the 'Silicon Glen' of Scotland is (comparatively) 'new'. Breheny, Hall and Hart (1987) pointed out that the decaying urban areas of Britain were not only old but had also been single-industry towns or cities. Of the 20 poorest areas in the country, 18 had been dependent upon mining, steel or shipping. However, although most of the victims of industrial decay were in the north, the list included Penzance, Falmouth, Torquay, Deal and Great Yarmouth. The booming towns, in contrast, have not had to suffer the legacy of the Industrial Revolution. They include a high proportion of the market and outer-suburban towns of Britain and other towns (and cities) that have attracted high-technology industries and associated service industries.

Throughout Britain, intra-regional disparities are very evident. All regions are experiencing, to a greater or lesser extent, an urban–rural shift in employment away from the major cities (see Hudson and Williams, 1986), and all regions have areas of boom and low unemployment, and areas of decay and high unemployment – notably within the inner cities (see Gudgin and Schofield, 1987). It is sometimes argued, moreover, that the problems of interregional disparity are, in essence, the problems of the inner cities (particularly the accumulation of housing, health, education and job deprivation), and that the north just happens to contain more inner cities than the south (TCPA, 1987).

There is thus no interregional demarcation line separating the 'haves' and the 'have nots'. While the poor (recruited mainly from the ill-educated, the young, the unskilled, the elderly and the growing number of single parents) are concentrated both in the decaying urban areas of the north and in Inner London, the relatively affluent majority are to be found virtually everywhere: and those in the north, in some ways, enjoy a higher standard of living than those in the south (Gudgin and Schofield, 1987; Wilsher and Cassidy, 1987).

It could be argued that, if there is a north-south divide, it is not a division between an impoverished north and an affluent south but a divide between two different sets of problems. Whereas deindustrialization and relatively high unemployment might be the two most serious causes for concern in the north, in the South East business efficiency is undermined by the cost-inflationary effects of skill shortages, Green-Belt constraints on land supply, and road and rail congestion; at the same time, the quality of life is impared by crippling mortgage payments, long commuting hours, lower (net) disposable incomes and arguably a poorer environment (TCPA, 1987). London itself, however, suffers many of the problems of the north. Indeed, far from it being an island of prosperity (as many people believe), the capital has some of the highest levels of unemployment, poverty and deprivation in Britain (Association of London Authorities, 1986) – magnitudes of disadvantage that are even less tolerable than elsewhere because of the high cost of living in the capital and the juxtaposition of areas of abject poverty and considerable wealth.

In this chapter, an examination of a range of economic and social indicators is intended to cast doubts on the contention that there is a stark 'north-south divide', and to point the way to a major change in the rationale of regional policy.

Employment disparity – some contradictions

The north: employment expansion in the 1980s

Whereas employment in the northern regions in aggregate substantially decreased during the depression of the late 1970s and early 1980s (the number of jobs diminishing by over 1,100,000 in 1979 to 1981), it was evident that there were significant rates of increase in employment in many parts of Britain north of the Severn–Wash line during (and beyond) this period. Large areas of Scotland, for example, showed notable employment gains between 1979 and 1981. Within this period, the number of jobs in Grampian increased by 9.9 per cent, in Dumfries & Galloway by 5.4 per cent, in the Island Areas by 2.6 per cent and in the Highlands by 2 per cent (Townsend, 1986). Although some of this increase in employment (particularly in Grampian) was attributable to the development of the oil industry, most new employment was generated by small manufacturing firms with better-than-average manufacturing performance (Business Statistics Office, 1981; Fothergill and Gudgin, 1982; O'Farell, 1985; Townsend and Peck, 1985). Over a longer-term period, the oil industry has attracted resources and activity to the Highlands and Islands of Scotland and has stimulated infrastructure development such as highway improvement from Inverness to Edinburgh. In the Lowlands, smokestack industries have been gradually replaced by high-technology activity, notably in the field of electronics (which provided over 40,000 jobs in about 3,000 companies in the late 1980s) (Osmond, 1988). Since 1975, when it was established, the Scottish Development Agency (SDA) can take credit for attracting new employment into Scotland - by bringing together public- and private-sector initiatives and through its many other activities. As Cameron (1985) noted, the SDA was particularly effective in promoting the growth of new sectors, retaining the local ownership of key indigenous companies and in major firm-rescue operations.

The West Midlands was another region to experience a turnaround in employment in recent years. While the region lost 287,600 jobs in the period 1979-87, most of the decline in employment was in the early 1980s. Since then the level of unemployment (largely associated with job loss) has diminished from its peak of 14.2 per cent in 1983 to 9 per cent in 1988; and whereas manufacturing output (nationally) grew at an annual rate of 7 per cent (in the third quarter of 1988), in the West Midlands output in the electrical and instrument-engineering industries increased by 15.4 per cent, in the metal industries by 11.9 per cent and in car production by 7.9 per cent in the same period. Recovery was also marked by the decrease in availability of industrial floor-space. In 1982, there was a glut of 28.1 million ft^2 of unlet or unsold space in the region, but by the end of 1988 this had decreased to less than 10 million ft^2 (Williams, 1988). Financial institutions, moreover, increasingly provided loans to businesses for development rather than cash for restructuring, and of all regions the West Midlands became the most successful in attracting foreign investment. By the late 1980s, according to the West Midlands Industrial Development Association, the region was receiving 25 per cent of total foreign investment in Great Britain and benefiting through the associated creation of 80,000 jobs.

A more precise picture of the spatial distribution of employment growth can be seen by examining data at the level of the travel-to-work area (TTWA), rather than at the regional level. Townsend (1986) not only revealed that of the 380 TTWAs in Great Britain, 115 (or 30.3 per cent) had shown an increase in employment in the period June 1978 to September 1981, but also showed that more than 60 per cent of these 115 areas lay outside of the South East, South West and East Anglia. There were many concentrations of increased employment in each northern region of Britain, for example, in the TTWAs of Aberdeen, Dumfries, York, Chester and Stafford. Employment growth was particularly notable in Scotland, where 40 per cent of its 60 TTWAs experienced increases, while in the West Midlands, Wales and the North growth was spread over proportionately more TTWAs than in East Anglia (Table 4.1).

Table 4.1   Employment increase in travel-to-work areas, Great Britain, 1978-81

	Total number of travel-to-work areas	Number of travel-to-work areas with increasing employment	(%)
South East	51	26	51.0
Scotland	60	24	40.0
South West	56	18	32.1
West Midlands	27	8	29.6
Wales	40	11	27.5
North	22	6	27.2
East Anglia	27	7	25.9
East Midlands	34	6	17.6
Yorkshire & Humberside	34	5	14.7
North West	29	4	13.8
Great Britain	380	115	30.3

(*Source*: Townsend, 1986.) The location of employment growth after 1978: the surprising significance of dispersed centres, *Environment and Planning A, Vol 18.)*

In a more detailed spatial context – the level of the employment office area (EOA) – it is evident that there was an even higher proportion of areas showing employment growth. Townsend (1986) not only revealed that of the 852 EOAs in Britain, 289 (or 33.9 per cent) had shown an increase in employment (1978-81), but also reported that more than half of these 289 areas were distributed throughout the north.

Employment growth continued in many parts of the north throughout the 1980s. This was reflected by an increase in the volume of property development and soaring land costs. In contrast to the south, the cost of land in the north was low and attracted an increasing amount of development in the latter part of the decade. The diversion of demand from the south (and particularly from the South East) pulled up land values substantially in the more favoured areas of the north – the cost of house-building land in 1988, for example, rising by 163.8 per cent in the West Midlands, 127.1 per cent in the East Midlands, 109.1 per cent in Yorkshire & Humberside and 90.6 per cent in Wales, compared to an increase of 68.9 per cent in the South East and only 22.8 per cent in Outer London (McGhie, 1988). Industrial and commercial land costs also rose as the demand for sites increased. In Sheffield, for example, £7 billion of large-scale development commenced in 1989, in Leeds and Manchester respectively £4 billion and £1.5 billion of development was scheduled to be undertaken in the early 1990s, while Merseyside intended to invest over £400 million in dockland redevelopment in the same period.

By the late 1980s it was becoming apparent that there were increasing shortages of land for development in many parts of the north, particularly in the Midlands and in Yorkshire & Humberside. Several local authorities (for example, Birmingham, Solihull, Coventry, Barnsley and Rotherham) seemed only too willing to consider sacrificing Green-Belt land in return for jobs and investment on the edge of their run-down industrial areas, whereas conservation groups (such as the Council for the Preservation of Rural England) normally favoured the renewal of derelict industrial land (Brough and Palmer, 1988).

The south – transformation or stagnation?

Throughout the 'Long Boom' years, economic growth in the South East was an important factor in regional planning (Keeble, 1980a; Damesick, 1982). 'Carrot-and-stick' policies designed to divert capital and hence jobs to areas of high unemployment in the north were dependent very largely on a buoyant economy in the South East (see Chapter 3). Since the late 1970s, however, the notion that the north can be, in effect, dependent on the South East 'is a dangerous oversimplification. It recalls the view of regional planning taken in the 1950s and 1960s, when London was a dynamo, generating ... employment that could contribute to the revival of other regions' (Wood, 1987, p. 64). During the thirty years after the Second World War, the north-south divide was seen essentially as a division between the South East and the rest of Britain (although it was acknowledged that the Midlands had many of the growth characteristics of the South East). By the 1980s, however, two important changes had taken place in the south that should have substantially modified this view. First, the ripple effect of prosperity had spread out from the South East to East Anglia and much of the South West to form the 'Greater South East' (Regional Studies Association, 1983); and, second, the local economy of many areas in the 'traditional South East', such as south Essex, Portsmouth, some coastal resorts, the new towns and north Kent began to stagnate or decline.

In the case of the new towns, an early post-war specialization in manufacturing proved to be a weakness in their economic structure, 'while in the Medway towns [of north Kent] the closure of the naval dockyards ... produced local economic problems reminiscent of those of northern industrial centres' (Damesick, 1987, p. 27). Whereas in the 1960s and 1970s, regional policy imposed restrain on indigenous growth in the South East, with frustrated and inefficient firms closing down rather than making a move to the north (Lock, 1988), in the early 1980s, 'in some parts of the South East ... neither location in the region nor proximity to London protected local manufacturing employment from the effects of the recession' (Wood, 1987, p. 66). Employment decline in manufacturing continued throughout the 1980s (most notably in the motor vehicle, food and drink, oil and chemicals, engineering and furniture-making industries). As Table 4.2 shows, over half a million jobs were lost in manufacturing in the South East (1979–87) – the region accounting for almost a quarter of the overall shake-out in manufacturing employment in the UK.

It has been generally assumed that in the early 1980s the north suffered a much greater decline in employment than the south. However, although the

Table 4.2   Regional share of employment decline in manufacturing, UK, 1979–87

	Employment change* 1979–87 ('000)	Share of total change (%)
South East	− 512.7	24.1
North West	− 360.6	17.0
West Midlands	− 287.6	13.5
Yorkshire & Humberside	− 259.2	12.2
Scotland	− 211.2	9.9
North	− 146.8	6.9
East Midlands	− 114.5	5.4
Wales	− 108.8	5.1
South West	− 73.2	3.4
Northern Ireland	− 44.2	2.1
East Anglia	− 1.3	0.1
UK	− 2,126.0	

Note
* Rounded to nearest thousand.

(*Source*: Department of Employment.)

Department of Employment (1987a) reported that 94 per cent of job losses were in the north and only 6 per cent in the south, these proportions represented net reductions and masked the very severe gross decline in employment (specifically in manufacturing) in the south. The increase in service employment in the South East (Table 4.3), although substantial, has not necessarily provided alternative jobs for displaced full-time male labour. The bulk of this employment was in routine catering, cleaning and clerical work (Stott, 1988) – much of it being low-paid, part-time employment for women (*New Society*, 1987).

Within the South East, the labour-market is becoming increasingly polarized.

Table 4.3   Trends in service employment, Great Britain, 1976–86

	Employment change 1976–86 ('000)	Share of total increase 1976–86 (%)
South East	+ 626	53.6
(Greater London)	(+ 87)	(7.4)
South West	+ 487	41.7
West Midlands	+ 137	11.7
East Midlands	+ 122	10.4
East Anglia	+ 66	5.6
North	+ 44	3.8
Scotland	+ 33	2.8
North West	+ 20	1.7
Wales	− 7	—
Great Britain	+ 1,169	

(*Source*: Department of Employment, *British Labour Statistics*.)

A widening gap is emerging between those in permanent jobs (in, for example, business and financial services and public administration) and those employed in monotonous, insecure and low-paid jobs, such as part-timers, temporary workers and home workers, and a segment of the self-employed. Stott (1988) suggested that a very real 'south-south' divide is thus emerging in the labour-market. On the one hand, very high salaries can be earned in business and financial services, while on the other hand, 'one-third of all workers in the South East [earned less] ... than the Council of Europe's "decency minimum" of £115 per week' (*ibid.* p. 238). With the privatization of many public services, it is probable that the gap between the advantaged and disadvantaged in the labour-market will widen – with the less well-off having to suffer lower real wages, longer hours and worse conditions.

Greater London: a major cause for concern

From the 1960s to at least the mid-1980s, the UK faced massive deindustrialization, the economic collapse of its major cities and a sustained process of population and job decentralization. Deindustrialization was particularly marked in London and the major conurbations, which together lost 1.7 million (79 per cent) of the total national loss of 2.1 million manufacturing jobs in the period 1960-81 (Greater London Council, 1985a).

Table 4.4 shows that after a period of employment growth in the 1950s, Greater London subsequently suffered substantial job losses (particularly in manufacturing). By the mid-1980s, while London still dominated many activities, such as business and financial services, 'the economic prospects of many inner Londoners [were] ... no better than those of workers in other conurbations' (Wood, 1987, p. 64). Employment levels in every sector of London's economy were worse than the national average, with manufacturing employment declining faster and service industries growing more slowly than in the UK as a whole – the capital consequently losing 184,000 jobs in the period 1978-86 (Association of London Authorities, 1988).

Table 4.4 Annual average employment change in Greater London, 1951–84

	Manufacturing	Production industries other than manufacturing	Services	Total employment
	('000)	('000)	('000)	('000)
1951–61	+400	−1,400	+18,700	+17,700
1961–6	−30,700	+4,700	−11,700	−14,300
1966–71	−38,400	−13,200	−2,700	−54,300
1971–4	−49,200	−6,200	+24,600	−30,800
1974–8	−33,100	−5,100	−3,600	−41,800
1978–81	−24,500	−4,500	−5,800	−34,800
1981–4	−32,300	−7,900	+5,400	−33,900

(*Sources*: *Censuses of Population*, 1951-71; *Censuses of Employment*, 1971-81; Department of Employment quarterly estimates.)

Chronic employment decline and economic atrophy in the capital was partly attributable to the decentralization of employment and population. In the 1960s and 1970s, all the boroughs in the eastern half of London suffered high rates of job loss in manufacturing, employment in this sector declining by up to 60 per cent in Greenwich, Hackney, Lewisham, Newham and Tower Hamlets (Buck *et al.*, 1986). In the early 1980s, however, economic crisis spread beyond inner London westwards to Hayes and Hounslow, Park Royal and Southall as firms such as AEC, Firestone, Hoover, Macleans and Pyrene undertook plant closure, and Heathrow Airport shook out 'surplus' labour. Whereas firms in the growth industries of the 1930s had located in west and north-west London, areas of recent growth are some 30 or more miles further from the capital in the M3/M4 corridors. Within this 'sunrise belt', the new high-technology industries of the 1970s and 1980s are far less significant as sources of employment than their pre-war counterparts (Buck *et al.*, 1986). Due to high travel costs, long distances to work opportunities and a dependency on council housing, low-paid or unemployed labour remains trapped in Greater London: a metropolis devoid of new growth industry, and a city fraught with social tensions in its inner areas (*ibid.*).

In part, job creation in Greater London has been constrained by relatively high labour costs. The Department of Employment (1986) reported that the average weekly earnings in London were 11 per cent above the average for male manual workers in 1984, 17 per cent above for non-manual males and for manual women, and 20 per cent above for non-manual women. It is notable that these differentials in pay did not diminish during the period of chronic employment decline in the early 1980s. Skill shortages have also impeded the generation of employment. The Manpower Services Commission (1987) identified over 2,000 skills shortages in the London labour-market both in new technology and in a wide range of traditional occupations in manufacturing, and the commission was concerned that although there were 73,000 participants in government training schemes in the capital in 1986, the schemes were insufficient to satisfy the demand for high-quality training (Association of London Authorities, 1988). Clearly the decline in London's manufacturing base has not only reduced the total number of jobs in the capital in recent years but also, because of the erosion of skilled and semi-skilled employment, unskilled workers (instead of being able to gain appropriate skills) are increasingly being employed in lower-grade service jobs (Damesick, 1987).

Manufacturing decline

With the growth of mass production after the First World War, London became one of the major manufacturing centres in the UK. During the depression of the early 1930s, 'Greater London remained an island of growth and relative prosperity' (Buck *et al.*, 1986, p. 1), with unemployment at less than two-thirds of the national average and at half the level of the northern regions. By 1951 there were 1.5 million people working in manufacturing industry in the capital, with manufacturing employment continuing to grow in the metropolis both absolutely and relatively throughout the decade - a result of extensive investment in new capacity. According to Hall (1962), London was still a strong, modern and growing manufacturing centre in 1960, but thereafter, 'there was a strong shift in manufacturing jobs away from the country's major cities, with the rate of job loss being particularly rapid in the Greater London area' (Damesick, 1987, p. 20). While the decline in manufacturing in London may be seen as part of the general deindustrialization of the British economy (Massey and Meegan, 1978; Danson, Lever and Malcom, 1980; Keeble, 1980b; Elias and Keogh, 1982; Fothergill and Gudgin, 1982), it is notable that in the period 1966-74 manufacturing employment in Greater London declined at around three times the national rate of loss (Dennis, 1978). Over a long period (1961-81), there was a decline of over three-quarters-of-a-million manufacturing jobs in the capital, the number decreasing from 1,453,000 to 680,000 with manufacturing's share of London's employment falling from 32 to 19 per cent (Buck et al., 1986).

Manufacturing employment loss in the capital in the 1960s and 1970s was attributable to three inter-related processes. First, there was a very high degree of job decentralization particularly from west London during the 1960s (Keeble, 1968): a process compounded by the relocation of economic activity to the north in consequence of regional 'carrot-and-stick' policy (see Chapter 3). Second, because of cramped sites, poor transportation links and shortages of skilled labour, manufacturing firms in Greater London were increasingly at a disadvantage compared to their competitors located elsewhere in the south. Gripaios (1977) showed that two-thirds of manufacturing closures in inner south-east London were the result of company 'deaths' rather than relocations; Dennis (1980) reported that 47 per cent of all job losses in Greater London in the period 1966-74 were due to closure, most of which were unconnected to moves; and in a survey conducted by the Confederation of British Industry (1989), it was revealed that road congestion in the capital cost local businesses as much as £15 billion per annum - clearly a very substantial diseconomy of locating in Greater London. Third, and of at least equal significance, the restructuring of British industry had a particularly marked effect on manufacturing in the capital. Prais (1976) revealed that by 1970 the largest 100 manufacturing firms in the UK accounted for as much as 41 per cent of the national output of manufactured products (compared to only 16 per cent in 1909), while Hannah and Kay (1977) suggested that - in respect of multi-plant firms - increased industrial concentration was caused by acquisitions and mergers rather than by internal growth. Many independent and relatively small firms in inner London confronted with the increased scale of activity of their rivals, either went out of business voluntarily or were taken over and then closed. There was also an increased tendency for large firms to retain their control functions in London but to decentralize routine production activity and research and development to greenfield areas elsewhere in the south (Massey, 1984).

In the ten years after the 'Long Boom', Greater London lost over one-quarter-of-a-million manufacturing jobs, nearly half of the decline being during the period 1978-81. In percentage terms, London suffered a greater reduction in manufacturing employment than the rest of the South East and Great Britain as a

Table 4.5　Changes in manufacturing employment in the South East, 1975–84

	1975–8		1978–81		1982–4	
	('000)	(%)	('000)	(%)	('000)	(%)
Greater London	−67	−8.0	−119	−15.5	−66	−10
Rest of South East	+15	+1.4	−101	−9.2	−8	−1
Great Britain	−217	−3.0	−1,193	−16.8	n.a.	−6

(*Sources*: *Censuses of Employment; Employment Gazette, Historical Supplement,* no. 1, April 1985.) 1985.)

whole (Table 4.5), and by 1985 London's manufacturing workforce was around half its size at the start of the 1960s (Damesick, 1987).

Although there were cuts in capacity and reorganization of working practices involving little new investment during the recession of the early 1980s, the pace of decentralization was less than during the preceding two decades (Townsend and Peck, 1985). Nevertheless, it was apparent that there were still very substantial cost disadvantages of locating in London (Moore, Rhodes and Tyler, 1984). Industrial rents in west London were over 75 per cent higher than in most areas of the South East (Hillier Parker Research, 1982), while average weekly earnings were up to 20 per cent above the national average. By the mid-1980s these higher costs (together with the absence of room for development in Greater London) led to a further dispersal of manufacturing away from the capital. It was unfortunate for London that while dynamic firms tended to move, the less advanced and weaker firms remained.

Despite the movement of manufacturing firms out of London, the capital still suffers from a substantial shortage of suitable and affordable land for industrial development. With a major upturn in the economy, the industrial prospects of London will remain bleak (unless there is a reliance on small and medium firms with only modest site requirements). While there may be an adequate supply of available land throughout Britain (and particularly in the north) to match the overall demand for industrial sites in the short-to-medium term[1], not all of this land is 'in the right place, available at the right time, or of the right size and the right price' (see Fothergill, Kitson and Monk, 1987). This is particularly the case in London and where Green-Belt restrictions limit supply elsewhere in the South East.

Service employment: an asset or a liability?

Service activities dominate London's employment structure. Comprising 40 per cent of the South East's population in 1981, the capital contained as many as 53 per cent of its service jobs. However, although service employment increased its share of total employment in the capital from 58 to 75 per cent in the period 1961–81, in absolute terms the increase was pitiful. Between 1961 and 1981 the number of service jobs in London increased by only 45,000 (from 2,620,000 to 2,665,100), and in inner London service employment actually declined by up to 25 per cent.

Table 4.6   Employment change in Greater London, 1973–83

Sector	Numbers employed (1973)	(1983)	Change (No.)	(%)
Financial, professional and miscellaneous services	1,397,716	1,468,000	+ 70,284	+ 5
Public administration and defence	344,700	313,000	− 31,700	− 9
Distributive trades	528,939	459,000	− 69,939	− 17
Infrastructure*	672,901	525,000	− 147,901	− 22
Manufacturing	924,086	594,000	− 330,086	− 36
Total employment	3,872,739	3,366,000	− 506,739	− 13

Note
* Construction, gas, electricity, water, transport and communications.

(*Source*: Greater London Council, 1985a.)

Although nationally the growth in the number of service jobs broadly compensated for the decline in manufacturing employment, in Greater London employment in all sectors of the economy (except financial, business and miscellaneous services) contracted the period 1973–83 (Table 4.6). Over the years 1978–86, however, the number of jobs in service industries in Greater London increased dramatically, particularly in financial services, although the growth in service employment still failed to compensate for the massive reduction in employment in manufacturing and construction (Figure 4.1).

Notwithstanding the recent growth of service employment in London, there is little doubt that service sectors in the capital have been underperforming compared to their counterparts elsewhere in the UK. The Association of London Authorities (ALA) (1988) claimed that if employment in the capital's service industries had grown at the national service-sector rate, an extra 34,000 service jobs would have been created over the period 1978–86. Technological change in communications and administrative procedures, curbs in the growth of public-sector activity, and decentralization each constrained the expansion of employment opportunities in the capital. The ALA argued that the principal reason for underperformance was the enormous rise in land values. In the period 1978–86, the average price of land per hectare increased by 475 per cent in London, up to a level of 4.6 times the national average. At an average of £1.2 million per hectare in 1986, the price of land in London deterred small- and medium-sized businesses from locating in the capital and encouraged service employers in London to realize their property assets and relocate elsewhere in Britain. Thus, although the service sector is still the leading activity in the metropolis, the problems it faces suggest that it no longer offers 'a sound basis for employment growth, or even its retention in London' (Wood, 1987, p. 84) – regardless of the many proposals to undertake massive commercial developments in the 1990s, for example, at Canary Wharf and King's Cross.

Unemployment – a re-examination

In the UK, most social scientists and many politicians, employers and trade unionists rightly express considerable concern at the north-south disparity in unemployment. But unemployment data are often interpreted without regard to regional inequalities throughout the European Community (EC), while variations in unemployment within the different regions of the UK and absolute levels of unemployment (as opposed to percentage rates) are usually ignored.

Within the EC in 1987, unemployment rates ranged from only 2.7 per cent in

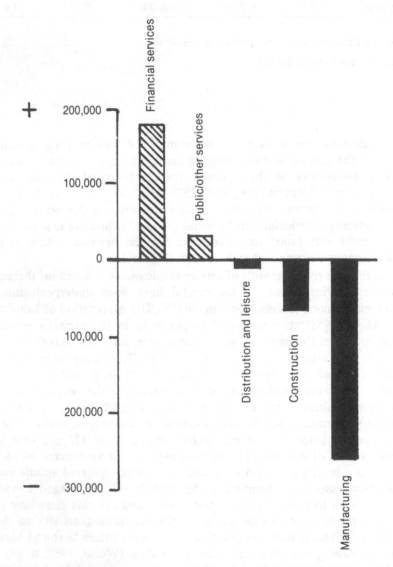

Figure 4.1   Employment change in Greater London, 1978–86 (Source: Greater London Council)

Luxemburg to 29.9 per cent in the Sur region of Spain. Interrregional disparities in the UK were therefore quite moderate within the context of these two extremes (Figure 4.2). Table 4.7 shows that, except for Northern Ireland, Scotland and the North, all the UK's regions had unemployment rates – within the second and third data quartiles – ranging from 8.1 per cent in East Anglia to 13.5 per cent in the North West.

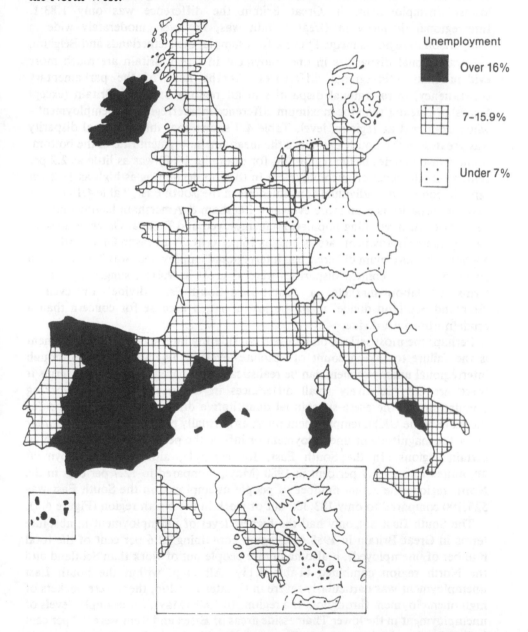

Unemployment

Over 16%

7–15.9%

Under 7%

Figure 4.2   Unemployment in the European Community, 1987 (Source: Statistical Office of the European Communities)

data quartiles – ranging from 8.1 per cent in East Anglia to 13.5 per cent in the North West.

Interrregional differences in unemployment rates in the UK (and particularly Great Britain) were also not excessive when compared to regional inequalities in West Germany and Spain (Table 4.8). For example, whereas in Italy there was a 3.46:1 difference between unemployment rates in the regions of highest and lowest unemployment, in Great Britain the difference was only 1.83:1. Interregional disparity in Great Britain was, however, moderately wide in comparison to regional inequalities in, for example, the Netherlands and Belgium.

Intra-regional disparities in unemployment in Great Britain are much more evident than interregional differences. At the level of the parliamentary constituency, intra-regional disparities in all regions of Great Britain (except Wales) are greater than the maximum difference in interregional unemployment – when examined at regional level. Table 4.9 shows that intra-regional disparity was greatest in the South East since the mean unemployment rate in the bottom-decile constituencies in the region (in, for example, 1988) was as little as 2.2 per cent, while the level of unemployment in the top decile was as high as 13.2 per cent. In comparing individual constituencies in the South East, Table 4.10 reveals that the disparity is even more evident – Chesham & Amersham having only 1.7 per cent of their working population unemployed and Bethnal Green & Stepney having an unemployment rate as high as 16 per cent. Whereas in Great Britain as a whole, the maximum difference in interregional employment was 2.35:1, within an intra-regional context disparities were as great as 9.41:1, suggesting that in terms of labour-markets, a 'South-East–South-East divide' or even a 'Scotland–Scotland divide' might be just as much a cause for concern than a crude north-south divide.

Perhaps the most serious omission in considering interregional unemployment is the failure to take account of absolute levels of unemployment. Although interregional unemployment can be realistically compared in percentage terms if there are only relatively small differences in the size of regional working populations, if there are substantial quantitative differences in population (as there are in the UK) unemployment rates, as normally presented, can easily *either* mask the magnitude or unemployment *or* inflate the perception of its severity in certain regions. In the South East, for example, although unemployment amounted to only 5.7 per cent in 1988 (May), compared to 12.7 per cent in the North region, the actual number of people unemployed in the South East was 523,100 compared to only 183,300 out of work in the North region (Figure 4.3).

The South East not only had the highest level of unemployment in absolute terms in Great Britain in 1988 (the region containing 22.6 per cent of the total number of unemployed), it also had more people out of work than Scotland and the North region combined (Table 4.11). Although within the South East unemployment was particularly severe in Greater London, there were pockets of high unemployment throughout the region. In 1988 (May), for example, levels of unemployment in the lower Thamesside areas of Essex and Kent were 8.7 per cent in Southend and 8.7 per cent in Thanet; in Buckinghamshire, unemployment in Luton was 9 per cent; in north Essex it reached 9.4 per cent in Harwich; on the

Table 4.7 Unemployment rates within the European Community, 1987

Rank	Region/nation	Unemployment rate (%)	Rank	Region/nation	Unemployment rate (%)
1	Sur	29.9	31	Abruzzi-Molise	9.9
2	Canarias	25.4		Oost-Nederland	9.9
3	Campania	21.2	32	Lazio	9.8
4	Este	20.6	33	Zuid-Nederland	9.6
5	*Northern Ireland*	18.9	34	Est	9.4
6	Sardegna	18.7		Vlaams gewest	9.4
7–8	Centro (Spain)	18.5	35	West-Nederland	9.3
	Noreste	18.5	36	Centre-Est	8.8
9	Ireland	18.1	37	*South West*	8.7
10	Madrid	16.2	38	Ile de France	8.6
11	Sicilia	16.0	39	Nord Ovest	8.5
12	Noroeste	15.5	40	Kentriki Ellada	8.3
	Sud	15.5	41	Nordrhein-Westfalen	8.2
13	*Scotland*	14.8		*South East*	8.2
14	*North*	14.7	42	Berlin-West	8.1
15	Region wallonne	14.4		*East Anglia*	8.1
16	Nord-Pas-de-Calais	14.0	43	Niedersachsen	7.9
17	*North West*	13.5	44	*Centro (Italy)*	7.7
18	Mediterranee	13.0	45	Schleswig-Holstein	7.3
19	*Wales*	12.9	46	Portugal	7.1
20	*Yorkshire & Humberside*	12.5	47	Nord-Est	7.0
21	*West Midlands*	12.4	48	Emilia-Romagna	6.7
22	Region bruxelloise	12.2	49	Voreia Ellada	6.4
23	Noord Nederland	11.6	50	Lombardia	6.1
	Bremen	11.6	51	Denmark	6.0
24	Bassin parisien	11.0	52	Rheinland-Pfalz	5.7
25	Ouest	10.9	53	Hessen	4.8
27	Hamburg	10.6	54	Anatolika kai notia nisia	4.6
28	Sud-Ouest	10.5	55	Bayern	4.4
29	Saarland	10.3	56	Baden-Württemberg	3.6
30	*East Midlands*	10.0	57	Luxemburg	2.7

Note UK regions shown in italics.
(*Source:* Central Statistical Office, *Regional Trends.*)

Table 4.8 National interregional differences in unemployment rates, European Community, 1987

	Highest rank region	Unemployment rate (%)	Lowest rank region	Unemployment rate (%)	Interregional difference (%)
Italy	Campania	21.1	Lombardia	6.1	346
West Germany	Bremen	11.6	Baden-Württemberg	3.6	322
UK	*Northern Ireland*	*18.9*	*East Anglia*	*8.1*	*233*
(Great Britain)	*(Scotland)*	*(14.8)*	*(East Anglia)*	*(8.1)*	*(183)*
Spain	Sur	29.9	Noroeste	15.5	193
Greece	Kentriki Ellada	8.3	Anatolika kai notia nisia	4.6	180
France	Nord-Pas-de-Calais	14.0	Ile de France	8.6	163
Belgium	Region wallonne	14.4	Vlaams gewest	9.4	153
Netherlands	Noord-Nederland	11.6	West-Nederland	9.3	125

(*Source:* Central Statistical Office, *Regional Trends.*)

Table 4.9 Interregional and intra-regional disparities in unemployment, top and bottom deciles, Great Britain, May 1988 (percentage)

	Unemployment rate	Mean unemployment rates		Differences between mean unemployment rates
		Top-decile constituencies	Bottom-decile constituencies	
South East	5.7	13.2	2.2	600
East Anglia	5.4	11.4	2.4	475
North West	11.3	22.7	4.9	463
South West	6.9	12.6	3.0	420
East Midlands	7.9	14.6	3.5	417
West Midlands	9.4	18.4	4.7	391
Yorkshire & Humberside	10.1	17.4	4.6	378
North	12.7	20.7	5.6	369
Scotland	11.9	20.6	6.1	337
Wales	11.2	14.8	7.2	206
Great Britain	8.4	12.7	5.4	235

Maximum difference in interregional unemployment   235

(*Source*: Department of Employment; Unemployment Unit.)

coast of Kent and Sussex it amounted to 8.9 per cent in Folkestone and 10.6 per cent in Brighton; while 9.6 per cent were unemployed on the Isle of Wight, 10 per cent in Southampton and 11.8 per cent in Portsmouth. Wood (1987) suggested that in the South East (outside Greater London), unemployment was concentrated *either* in those areas where employment in manufacturing and related services had been declining in the 1970s (for example, in lower Thamesside, Luton and Portsmouth), *or* where population growth had outstripped the generation of new jobs, notably in Milton Keynes - where, although unemployment at 7 per cent was above the national level (in May 1988), it was above the aggregate rate for the South East region.

Elsewhere in the south, unemployment above the national average occurred both in areas of industrial recession (specifically Norwich and Peterborough in East Anglia, and in Bristol in the South West) and in several coastal areas where diverse activities such as tourism, manufacturing, fishing and mining were in decline or had disappeared, for example, Great Yarmouth and Lowestoft in East Anglia, and Falmouth & Camborne, North Cornwall, St Ives, Truro, Plymouth, Torbay and Bournemouth in the South West. The rate of unemployment in these areas ranged from 8.5 per cent to 15.7 per cent in 1988 (May).

While there are thus many areas of high unemployment throughout the south outside Greater London, it is within the capital that the highest unemployment

Table 4.10 Interregional and intra-regional disparities in unemployment, top and bottom constituencies, Great Britain, May 1988 (percentages)

	Unemployment rate	Top constituency	Unemployment rate	Bottom constituency	Unemployment rate	Intra-regional differences
South East	5.7	Bethnal Green & Stepney	16.0	Chesham & Amersham	1.7	941
North West	11.3	Liverpool Riverside	28.6	Ribble Valley	3.8	753
North	12.7	Middlesbrough	24.6	Westmoreland & Lonsdale	3.4	723
East Midlands	7.9	Nottingham East	19.1	Harborough	2.9	655
East Anglia	5.4	Great Yarmouth	12.7	Cambridgeshire South East	2.1	605
West Midlands	9.4	Birmingham Small Heath	22.7	Stratford-upon-Avon	3.9	582
Yorkshire & Humberside	10.1	Sheffield Central	21.3	Skipton & Ripon	4.1	520
South West	6.9	Plymouth Drake	17.7	Dorset North	3.3	476
Scotland	11.9	Glasgow Central	22.4	Tweeddale, Ettrick & Lauderdale	5.1	439
Wales	11.2	Cardiff Central	15.5	Brecon & Radnor	7.0	221
Maximum difference in interregional unemployment						235

(*Source*: Department of Employment; Unemployment Unit.)

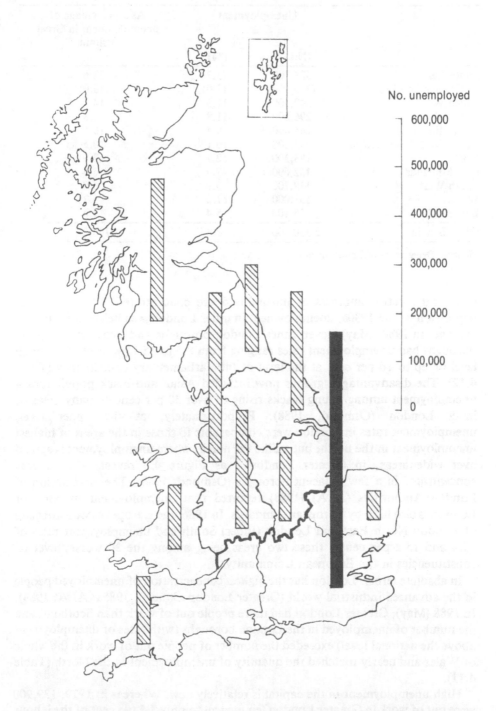

No. unemployed

600,000

500,000

400,000

300,000

200,000

100,000

0

Figure 4.3   Unemployment in Great Britain, 1988. Absolute Levels of Unemployment: The South East compared to other Regions of the UK. (Source: Department of Employment)

Table 4.11   Regional unemployment, Great Britain, May 1988

	Unemployment		As a percentage of unemployment in Great Britain
	(No.)	(%)	
South East	523,100	5.7	22.6
(Greater London)	(299,900)	(7.0)	(13.0)
North West	340,300	11.3	14.7
Scotland	296,800	11.9	12.8
West Midlands	244,800	9.4	10.6
Yorkshire & Humberside	242,100	10.1	10.5
North	183,300	12.7	7.9
East Midlands	152,600	7.9	6.6
South West	139,700	6.9	6.0
Wales	133,000	11.2	5.7
East Anglia	55,100	5.4	2.4
Great Britain	2,310,700	8.4	100.0

(*Source*: Department of Employment.)

rates and greatest numbers of unemployed are concentrated. Whereas in the depression of the 1930s, unemployment in inner London was below the national average, in 1988 (May) eleven Inner London boroughs and two Outer London boroughs had unemployment rates ranging from 8.5 to 15.1 per cent at borough level or up to 16 per cent at the level of the parliamentary constituency (Table 4.12). The disadvantaged groups now included Asian and black populations – unemployment among young blacks rising to over 30 per cent in many areas of inner London (Osmond, 1988). Proportionately, at the upper level, unemployment rates in London were comparable to those in the areas of highest unemployment in the north, but unlike the north, where unemployment is spread over wide areas, in Greater London – as Figure 4.4 reveals – it is very concentrated in a few adjacent boroughs (Osmond, 1988). The Association of London Authorities (ALA) (1988) reported that unemployment in parts of London is also high by European standards. In 1986, the European constituencies of London North East and London (Inner) South had unemployment rates of 15.4 and 15.2 per cent – these two areas being among the 30 worst 'level II' constituencies in the European Community.

In absolute terms, London has the highest concentration of unemployed people in the advanced industrial world (Greater London Council, 1985a; ALA, 1988). In 1988 (May), Greater London had more people out of work than Scotland, and the number of unemployed in the London boroughs (with rates of unemployment above the national level) exceeded the number of people out of work in the whole of Wales and nearly matched the quantity of unemployment in the North (Table 4.11).

High unemployment in the capital is relatively new. Whereas in 1979, 139,900 were out of work in Greater London (equivalent to only 3.5 per cent of the labour force), in 1988 (May) unemployment had more than doubled to 299,000 (or a rate

Table 4.12 Boroughs and constituencies in Greater London with disproportionately high levels of unemployment, May 1988

Borough	Unemployment (No.)	(%)	Constituency	Unemployment (%)
Tower Hamlets	13,520	15.1	Bethnal Green & Stepney	16.0
			Bow & Poplar	14.2
Hackney	16,837	14.8	Hackney South	15.5
			Hackney North	14.1
Islington	13,349	13.1	Islington North	13.8
			Islington South & Finsbury	12.4
Southwark	17,163	12.9	Peckham	15.4
			Southwark & Bermondsey	14.9
			Dulwich	8.4
Lambeth	19,661	12.4	Vauxhall	13.9
			Norwood	13.5
			Streatham	9.6
Haringey	14,356	11.2	Tottenham	13.7
			Hornsey & Wood Green	9.0
Camden	11,661	10.8	Holborn & St Pancras	13.2
			Hampstead & Highgate	8.4
Hammersmith & Fulham	9,774	10.2	Hammersmith	11.3
			Fulham	9.2
Newham	13,357	10.2	Newham North West	10.9
			Newham North East	9.1
Lewisham	14,260	9.9	Lewisham Deptford	13.3
			Lewisham South	11.0
			Lewisham North	8.6
Brent	13,889	8.6	Brent East	10.7
			Brent South	10.1
Greenwich	11,020	8.6	Woolwich	10.6
			Greenwich	9.2
Westminster	9,675	8.5	Westminster North	10.3
High unemployment boroughs and constituencies	178,519	11.3		11.8
Great Britain		8.4		

(*Source*: Unemployment Unit.)

of 7 per cent). In the North region, by contrast, unemployment had increased by only 6.2 per cent in the same period.

Although in broad regional terms, long-term unemployment represents a much higher proportion of total unemployment in the northern regions than in the South East (see Table 2.9), in Greater London the long-term unemployment rate had dramatically increased in recent years, for example, from 15.8 per cent in

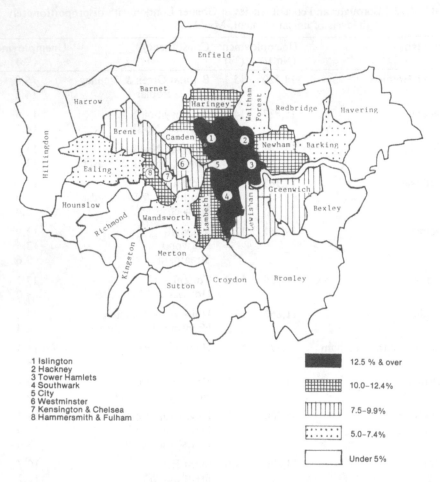

1 Islington
2 Hackney
3 Tower Hamlets
4 Southwark
5 City
6 Westminster
7 Kensington & Chelsea
8 Hammersmith & Fulham

12.5 % & over

10.0–12.4%

7.5–9.9%

5.0–7.4%

Under 5%

Figure 4.4   Unemployment in Greater London, 1988. (Source: Unemployment Unit)

1980 (October) to 41.6 per cent in 1987 (October). In absolute terms, this amounted to an increase in the number of long-term unemployed from 32,142 to 141,948 – the latter figure being larger than the total amount of unemployment in Greater London in 1978 (ALA, 1988). Among the under-25s in the capital, the long-term unemployment rate was as high as 26.4 per cent: indicating an urgent need for appropriate training and retraining.

Clearly, much of London is appallingly depressed. While a great deal of wealth and employment is created in many parts of the capital, little trickles down to the disadvantaged in inner London (TCPA, 1987). There is thus a growing divide between the 'haves' and the 'have nots', probably to a greater extent than in areas of high unemployment in the north.

But why is unemployment so high in inner London? There are a number of interconnected reasons. First, since the late nineteenth century, there has been a tendency for both industry and more highly qualified labour to decentralize to secure space for expansion and a desirable residential environment. In inner London, this has resulted in a mismatch between the supply of labour that remains (which is largely unskilled) and a greatly diminished number of jobs. Second, although there were employment losses on a massive scale in London throughout most of the 1960s and 1970s, the out-migration of labour in search of employment opportunities and the creation of new jobs in the capital ensured that unemployment rates in London remained below the national level throughout this period. After 1979, however, absolute levels of unemployment in inner London soared, partly because of the impact of deflationary macroeconomic policy (particularly in the period to 1983) and partly due to in-migration of unemployed labour from areas of low opportunity elsewhere in Britain (Buck *et al.*, 1986). Third, it can be demonstrated that there is a relationship between housing tenure and the incidence of unemployment. In areas where owner-occupation is the dominant tenure (the outer-London suburbs), unemployment is relatively low (generally below the national average). It can consequently be suggested that, other things being equal, owner-occupiers, mortgaged up to the hilt and under current interest-relief provisions, will inevitably have a stronger work-incentive than tenants, and be far less willing to risk employment in casual activity (McCormick, 1983). Owner-occupied housing, moreover, tends to attract people already qualified and already employed. In inner London (as in most inner cities in Britain), public-sector housing is the principal tenure and, since it is the cheapest form of housing, accommodates a disproportionate number of unemployed (Metcalf and Richardson, 1976). In inner west London and Brent, moreover, there is a relatively high concentration of private rented accommodation that attracts young people from elsewhere in Britain, adding directly or indirectly to the level of unemployment. However, although the problems of inner London occur very largely because the disadvantaged are forced to live in the inner city since that is where low-cost housing is chiefly found (Metcalf and Richardson, 1976), it is precisely because inner London contains a surplus supply of unskilled workers in relation to demand that unemployment continues to remain high in the inner areas of the capital.

Inequalities in production and expenditure – a cautionary view

Undoubtedly, there are interregional inequalities in production and levels of personal expenditure in the UK, as is shown in Chapter 2. However, disparities are less evident when gross domestic product (GDP) per capita is examined both within the context of the European Community (EC) and at the level of the county. When trends in the retail trade are also considered, the south rather than the north appeared to be disadvantaged in the late 1980s.

Table 4.13 Gross domestic product per capita within the European Community, 1986

Rank	Region/nation	Gross domestic product per capita (EC = 100)	Rank	Region/nation	Gross domestic product per capita (EC = 100)
1	Hamburg	189	23	*South West*	97
2	Ile de France	163		*Yorkshire & Humberside*	97
3	Region bruxelloise	155	24	Mediterranee	96
4	Noord Nederland	150		Sud-Ouest	96
5	Bremen	148	25	Schleswig-Holstein	95
6	Lombardia	134		*West Midlands*	95
7	Emilia-Romagna	131	26	Nord-Pas-de-Calais	94
8	Hessen	129	27	Zuid-Nederland	93
	Nord Ovest	129	28	*North*	92
9	Berlin-West	128		Ouest	92
10	Luxemburg	127	29	*Wales*	87
11	*South East*	121	30	Noreste	86
12	Baden-Württemberg	120		Oost-Nederland	86
13	Denmark	117	31	Region wallonne	84
14	Bayern	114	32	Albruzzi-Molise	83
15	Centro (Italy)	111		Madrid	83
	Nordrhein-Westfalen	111	33	*Northern Ireland*	81
	Lazio	110	34	Este	80
16	Nord Est	110	35	Sardegna	77
	West-Nederland	110	36	Campania	72
	Saarland	110	37	Sicilia	71
17	Est	105		Sud	71
18	Centre-Est	104	38	Noroeste	69
19	Bassin parisien	102	39	Canarias	66
	East Anglia	102	40	Centro (Spain)	64
	Vlaams gewest	102		Ireland	64
20	Rheinland-Pfalz	101	41	Sur	57
	East Midlands	101	42	Portugal	53
21	*Scotland*	99	—	Anatolika kai notia nisia	n.a.
22	Niedersachsen	98	—	Kentrike Ellada	n.a.
	North West	98	—	Voreia Ellada	n.a.

Note UK regions shown in italics.
(*Source:* Statistical Office of the European Communities.)

Although regional GDPs per capita within the EC in 1986 ranged from as little as 53 in Portugal to 189 in Hamburg (taking the average for the European Community as 100), in the UK per-capita GDP variations were notably less – ranging from 81 in Northern Ireland to 121 in the South East. Nine of the eleven regions of the UK had per-capita GDPs within the second- and third-data quartiles (Table 4.13). Thus, in contrast to the wide range of living standards throughout the EC (as measured by per-capita GDPs), the standard of living throughout most of the UK was comparatively uniform (Figure 4.5).

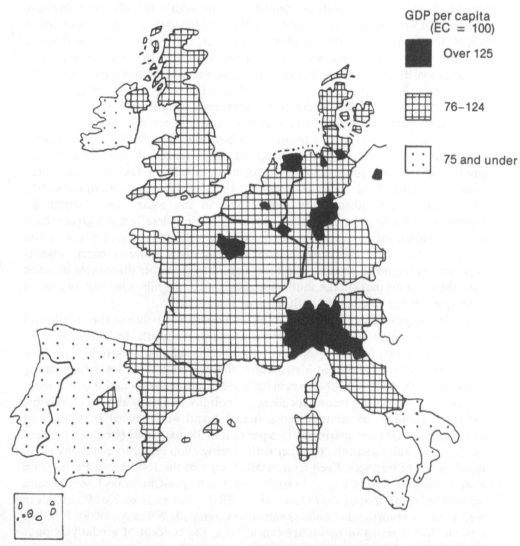

GDP per capita
(EC = 100)

■ Over 125

▦ 76–124

▦ 75 and under

Figure 4.5 Gross domestic products per capita in the European Community, 1986 (Source: Statistical Office of the European Communities)

It is also evident that among most member states of the EC, interregional differences in GDP per capita are wider than in the UK. Table 4.14 reveals that whereas there was a 149-per-cent difference between the most productive and least productive region in the UK in 1986, in West Germany and Italy differences were as much as 199 and 189 per cent respectively. Therefore in terms of regional GDPs per capita, the UK could be said to have the most spatially balanced economy in the EC.

An examination of GDPs per capita at the county level might also dispel certain myths. Table 4.15 shows that while eight of the top-quartile counties in 1984 were 'southern', as many as seven were 'northern' – with Grampian and Cheshire ranking second and fifth in the whole of Great Britain. The mean GDP per capita of the top-quartile counties in the south (£5,228) was broadly comparable with the mean per-capita GDP of the top-quartile northern counties (£5,045). Within the top-two quartiles, moreover, there were equal numbers of southern and northern counties. It is apparent, therefore, that whereas there is a spatial disparity between the distribution of the least prosperous counties (clearly a manifestation of the north-south divide), GDP data shows no significant 'divide' in the distribution of the more prosperous counties of Britain.

In recent years there has been a general convergence of wage levels, with average wages rising relatively in the north but declining relatively in the south. This was a reflection of the increased importance of low-wage service employment in the south and the associated growth in female employment (Gudgin and Schofield, 1987). Throughout the 1980s, house prices in the north, like-for-like, were substantially lower than in the south (see Chapter 2). Interregional house-price disparities, moreover, were generally much greater than regional variations in wages – partly because nationally uniform pay scales became increasingly the norm. After meeting their housing costs, owner-occupiers in the north therefore enjoyed significantly higher disposable incomes than their counterparts in the south mortgaged up to the hilt. This had important consequences for retailing in both parts of the country.

While it appeared in the south that the long consumer-boom of the middle and late 1980s was over, spending continued apace in the north. In December 1988 there were reports, for example, from both the Eldon Square centre in Newcastle upon Tyne and the Gateshead Metrocentre, that the retail trade was still booming (Elliott, 1988), while in Rotherham in early 1989, a branch of a major travel agent – Pickfords – reported record spending on holiday bookings abroad (Williams, 1989). In London, by contrast, consumer demand was reduced as a result of mortgage interest rates soaring to 13.5 per cent in January 1989 (from 9.5 per cent the previous July) and this, together with soaring shop rents, forced many small retailers out of business. Even large retailers such as the John Lewis Partnership were not immune from a drop in trade – their 1988 pre-Christmas boom in sales was less than half that of 1987 (Jaskowiak, 1989) – while a branch of Pickfords in west London reported that holiday sales had stagnated (Williams, 1989). It is thus evident that in terms of consumer expenditure, the concept of a relatively poor north and an affluent south is somewhat over simple.

Table 4.14 National interregional differences in gross domestic product per capita, European Community, 1986

	Highest-rank region	Gross domestic product per capita (EC = 100)	Lowest-rank region	Gross domestic product per capita (EC = 100)	Interregional differences (%)
West Germany	Hamburg	189	Schleswig-Holstein	95	199
Italy	Lombardia	134	Sicilia/Sud	71	189
Belgium	Region bruxelloise	155	Region wallonne	84	185
France	Ile de France	163	Ouest	92	177
Netherlands	Noord-Nederland	150	Oost-Nederland	86	174
Spain	Noreste	86	Sur	57	151
UK	*South East*	*121*	*Northern Ireland*	*81*	*149*
(Great Britain)	*(South East)*	*(121)*	*(Wales)*	*(87)*	*(139)*

(*Source*: Statistical Office of the European Communities.)

Table 4.15   Gross domestic product per capita, counties of Great Britain, 1984

Rank	'Southern' counties	GDP per capita (£)	Rank	'Northern' counties	GDP per capita (£)
1	Greater London	6,584			
			2	Grampian*	5,964
3	Berkshire	5,495			
4	Cambridgeshire	5,347			
			5	Cheshire	5,128
6	Hertfordshire	5,100			
			7	South Glamorgan	4,987
			8	Leicestershire	4,888
9	Surrey	4,882			
10	Wiltshire	4,826			
11	Buckinghamshire	4,824			
			12	Cumbria	4,805
			13	Lothian*	4,785
14	Bedfordshire	4,762			
			15	North Yorkshire	4,756
16	Avon	4,749			
17	Hampshire	4,717			
			18	Fife*	4,591
19	Oxfordshire	4,587			
20	Gloucestershire	4,573			
			21	Nottinghamshire	4,527
			22	West Midlands	4,492
23	Suffolk	4,477			
			24	Northamptonshire	4,473
			25	Cleveland	4,469
26	West Sussex	4,439			
27	Norfolk	4,436			
			28	Greater Manchester	4,355
			29	Derbyshire	4,334
			30	Highland*	4,307
			31	Lincolnshire	4,227
			32	Tyne & Wear	4,220
			33	West Yorkshire	4,211
34	Essex	4,147			
35	Devon	4,146			
			36	Lancashire	4,136
37	Dorset	4,126			
			38	Humberside	4,115
			39	Tayside*	4,087
			40	Strathclyde*	4,073
			41	Warwickshire	4,066
			42	Central (Scotland)*	4,045
			43	Dumfries & Galloway*	4,021
			44	Merseyside	3,928
			45	Staffordshire	3,922
46	East Sussex	3,908			
47	Kent	3,893	47	South Yorkshire	3,893
			49	Dyfed	3,857
			49	Gwent	3,857

Table 4.15   (*Cont.*)

Rank	'Southern' counties	GDP per capita (£)	Rank	'Northern' counties	GDP per capita (£)
			51	Hereford & Worcester	3,744
			52	West Glamorgan	3,728
			53	Northumberland	3,706
			54	Clwyd	3,705
55	Cornwall	3,619			
			56	Shropshire	3,599
			57	Co. Durham	3,567
			58	Gwynedd	3,435
59	Somerset	3,393			
			60	Mid Glamorgan	3,208

Note
* Local authority regions in Scotland.
(*Source*: Central Statistical Office, *Regional Trends*.)

The decentralization of population and employment

Population

Interregional migration, from the north to the south of Britain, is substantially overlaid by comparatively short-distance migration both within and between regions. Prior to an examination of intra-regional migration, however, it is necessary to consider population change at the urban level.

While the south and particularly the South East is dominated by one city, Greater London, in contrast to the north, which is polycentred, the major urban areas of England in terms of their social, economic and physical structure have much in common. The South East, for example, is divided into Inner and Outer London, the Outer Metropolitan Area and the Outer South East (see Hall *et al.*, 1987), and the north can be divided in a similar way into the Inner Urban North,

Table 4.16   Decentralization of population in the South East and North of England

	Population change (%)	
	1971–81	1981–6
Inner London	− 17.6	− 3.0
Inner Urban North	− 8.3	− 2.6
Outer London	− 4.6	0.4
Outer Urban North	0.4	− 1.2
Outer Metropolitan Area	4.9	2.4
Suburban North	4.0	1.4
Outer South East	8.2	9.3
Rural and Resort North	5.1	2.1

(*Source*: 1981 *Census*; Office of Population Censuses and Surveys, *OPCS Monitor*.)

Outer Urban North, Suburban North, and Rural and Resort North (see Webber and Craig, 1976; Champion, 1987). Patterns of population change within these broadly common structures are also alike. Table 4.16 shows that both Inner London and the Inner Urban North lost population in the periods 1971–81 and 1981–6, and population decline in Outer London (1971–81) was followed by decline in the Outer Urban North (1981–6). In the Outer Metropolitan Area and Suburban North, and in the Outer South East and Rural and Resort North, population increase occurred in the period 1971–81 and 1981–6. It is thus 'clear that decline in the north's major centres has been paralleled by growth in the outer areas, just as it has been in the South East' (Breheny, Hall and Hart, 1987, p. 25).

Table 4.17 shows that in Greater London and in each of the English conurbations out-migration greatly exceeded natural change (the relationship between birth and death rates), and also indicates that out-migration from the conurbations was (in percentage terms) greater than out-migration from the regions (except in the South East where migration into the region was even greater than the very substantial out-migration from Greater London).

A more detailed examination of population change not only indicates that, in addition to the migration trends outlined above, virtually all the metropolitan areas of the English conurbations (and inner and outer London in aggregate) experienced net out-migration in the period 1981–7, but also shows that within each region there was both a number of declining towns with high rates of net out-migration and growing towns with high rates of net in-migration (Table 4.18).

Table 4.17 Regional and conurbation population change, England, 1981–7

	Components of change, 1981–7		Change, 1981–7		Population, June 1987
	Natural change	Migration	('000)	(%)	
South East region	210.1	97.1	307.2	1.8	17,317.6
Greater London	115.9	− 151.2	− 35.3	− 0.5	6,770.4
West Midland region	72.9	− 61.4	11.5	− 0.2	5,197.7
West Midlands	46.3	− 95.2	− 48.9	− 1.9	2,624.3
North West region	34.8	− 124.2	− 89.4	− 1.4	6,370.0
Greater Manchester	21.8	− 60.9	− 39.1	− 1.5	2,580.1
Merseyside	8.1	− 73.2	− 65.7	− 4.4	1,456.8
Yorkshire & Humberside region	25.3	− 43.5	− 18.2	− 0.4	4,900.2
South Yorkshire	5.5	− 26.9	− 21.4	− 1.7	1,295.6
West Yorkshire	19.0	− 33.4	− 13.9	− 0.7	2,052.4
North region	8.7	− 49.6	− 40.8	− 1.3	3,076.8
Tyne & Wear	− 0.9	− 20.3	− 19.4	− 1.7	1,135.8

(*Source*: Office of Population Censuses and Surveys, *OPCS Monitor*.)

Table 4.18   Population change, English regions and metropolitan areas, 1981-7

	Components of change, 1981-7		Change, 1981-7	
	Natural change	Migration	('000)	(%)
South East region	210.1	97.1	307.2	1.8
Greater London	115.9	− 151.2	− 35.3	− 0.5
Inner London	53.9	− 92.0	− 38.0	− 1.5
Outer London	62.0	− 59.2	2.8	0.1
Remainder of South East	94.2	248.3	342.4	3.3
Declining towns:				
Brighton	− 2.8	− 5.7	− 8.5	− 5.7
Southampton	2.4	− 13.3	− 10.8	− 5.3
Reading	3.7	− 7.3	− 3.6	− 2.6
Portsmouth	0.2	− 4.8	− 4.6	− 2.4
Oxford	1.0	− 1.4	− 0.4	− 0.3
Growing towns:				
Milton Keynes	9.3	35.5	44.8	32.9
Wokingham	4.3	24.4	28.6	23.2
Crawley	3.1	8.6	11.6	15.8
Bracknell	4.0	6.3	10.3	11.8
Basingstoke	4.2	3.2	7.4	5.6
West Midlands region	72.9	− 61.4	11.5	− 0.2
West Midlands	46.3	− 95.2	− 48.9	− 1.9
Sandwell	3.4	− 14.8	− 11.4	− 3.7
Coventry	6.5	− 17.0	− 10.5	− 3.3
Wolverhampton	3.2	− 9.3	− 6.1	− 2.4
Birmingham	21.8	− 44.4	− 22.6	− 2.2
Walsall	4.4	− 10.3	− 5.9	− 2.2
Dudley	3.3	− 1.4	1.8	0.6
Solihull	3.8	2.0	5.8	3.0
Remainder of West Midlands	26.7	33.8	60.3	2.4
Declining towns:				
Rugby	0.8	− 3.0	− 2.3	− 2.6
Stoke-on-Trent	0.8	− 6.4	− 5.6	− 2.2
Newcastle under Lyme	0.3	− 2.7	− 2.4	− 2.1
Growing towns:				
Redditch	4.4	4.5	8.9	12.7
Stratford-upon-Avon	0.5	7.0	6.5	6.4
Worcester	0.8	3.2	4.0	5.2
North West region	34.8	− 124.2	− 89.4	− 1.4
Greater Manchester	21.8	− 60.9	− 39.1	− 1.5
Salford	− 0.8	− 8.5	− 9.3	− 3.8
Manchester	4.2	− 16.8	− 12.6	− 2.7
Trafford	1.6	− 7.2	− 5.6	− 2.5
Bury	0.9	− 3.9	− 3.0	− 1.7
Tameside	1.7	− 4.3	− 2.5	− 1.2
Wigan	3.0	− 6.2	− 3.1	− 1.0

Table 4.18   (*Cont.*)

	Components of change, 1981-7		Change, 1981-7	
	Natural change	Migration	('000)	(%)
Oldham	2.6	-4.5	-1.9	-0.9
Rochdale	4.0	-5.5	-1.5	-0.8
Bolton	3.0	-3.0	-1.0	0.0
Merseyside	8.1	-73.2	-65.7	-4.4
Liverpool	1.8	-42.5	-40.7	-8.3
Knowsley	6.9	-19.5	-12.5	-7.6
Wirral	0.6	-6.4	-5.8	-1.7
St Helens	0.5	-3.4	-2.9	-1.5
Sefton	-1.7	-1.4	-3.1	-1.1
Remainder of North West	4.9	9.9	14.8	0.6
Declining towns:				
Burnley	0.3	-9.5	-9.2	-9.9
Blackburn	2.6	-8.3	-8.7	-4.0
Ellesmere Port	2.3	-5.5	-3.1	-3.9
Growing towns:				
Warrington	2.2	11.7	13.9	8.1
Chorley	1.2	3.9	5.1	5.3
Lancaster	-2.8	7.9	5.1	4.0
Yorkshire & Humberside region	25.3	-43.5	-18.2	-0.4
South Yorkshire	5.5	-26.9	-21.4	-1.7
Sheffield	-5.5	-9.8	-15.3	-2.8
Barnsley	1.5	-5.8	-4.3	-1.9
Rotherham	4.3	-5.3	-1.0	-0.4
Doncaster	5.2	-6.0	-0.8	-0.3
West Yorkshire	19.0	-33.4	-13.9	-0.7
Leeds	2.7	-11.3	-8.6	-1.3
Wakefield	2.9	-6.8	-3.9	-1.2
Bradford	10.2	-12.6	-2.4	-0.5
Kirklees	3.3	-4.7	-1.4	-0.4
Calderdale	-0.1	2.1	1.9	1.0
Remainder of Yorkshire & Humberside	0.9	16.8	17.6	1.1
Declining towns:				
Hull	4.4	-25.4	-21.0	-7.9
Scunthorpe	1.3	-6.1	-4.8	-7.5
Great Grimsby	2.1	-5.0	-2.8	-3.1
Growing districts:				
Richmondshire	1.0	5.2	6.2	13.9
Selby	0.4	8.7	9.1	11.1
East Yorkshire	-1.8	7.8	6.1	7.9
North region	8.7	-49.6	-40.8	-1.3
Tyne & Wear	-0.9	-20.3	-19.4	-1.7

Table 4.18   *(Cont.)*

| | Components of change, 1981-7 | | Change, 1981-7 | |
	Natural change	Migration	('000)	(%)
South Tyneside	−0.9	−4.9	−5.7	−3.6
Gateshead	−0.7	−5.6	−6.3	−3.0
North Tyneside	−1.3	−4.4	−5.7	−2.9
Newcastle upon Tyne	−1.1	−0.3	−4.6	−0.5
Sunderland	4.9	5.1	0.2	−0.1
Remainder of North	−7.8	−29.3	−21.4	−1.1
Declining towns:				
Easington	0.4	−6.1	−5.7	−5.6
Sedgefield	0.6	−5.7	−5.1	−5.5
Hartlepool	0.9	−6.1	−5.1	−5.5
Growing districts:				
Eden	−0.6	2.8	2.2	5.1
South Lakeland	−2.3	6.4	4.1	4.3
Berwick-upon-Tweed	−0.7	1.6	0.9	3.4

(*Source*: Office of Population Censuses and Surveys, *OPCS Monitor*.)

In discussing the pace at which deurbanization is superseding suburbanization, Hall (1987a) examined population change at the level of the county. He showed that the following counties decreased in population (in descending order) by 0.3 to 0.8 per cent per annum (1981-4): Humberside, South Yorkshire, Tyne & Wear, Co. Durham, Gwent, Greater Manchester, the West Midlands, Mid Glamorgan, West Glamorgan, Merseyside, Cleveland and Inner London; whereas (in increasing order) Devon, Suffolk, Norfolk, Cambridgeshire, Buckinghamshire, North Yorkshire, Somerset, West Sussex, Cornwall, the Isle of Wight, East Sussex and Dorset grew by 0.7 to 1.5 per cent per annum (over the same period). As Hall suggested, all the losers were essentially areas of old industry or port activity, while the gainers were largely rural or coastal.

In the more accessible rural areas the in-migration of population from cities has recently begun to exceed depopulations. With regard to the 33 fastest-growing rural areas in England, Cross (1988) showed that 97 per cent of their population growth of 529,000 (1971-86) was attributable to in-migration, and that up to 70 per cent of gains were due to long-distance migration between regions. Although most of the fastest-growing areas were scattered throughout the south - for example, Wimborne (Dorset), Huntingdon (Cambs.), Caradon (near Plymouth), North Cornwall, Forest Heath and Babergh (Suffolk), and Breckland (Norfolk) - a number of growth areas were situated in parts of the north, such as Radnor (Powys), Ryedale (N. Yorks.), Selby, East Yorkshire and Holderness (Lincs.). It is unlikely, moreover, that the drift of population from urban to rural areas will wane in the foreseeable future. Clarke (1986) predicted that one-tenth of Britain's city dwellers (about 5 million people, for example) will migrate to the country within the next 30-40 years - the exodus from the cities resulting from the improvement of travel facilities (for example, the extension of high-speed rail links), the decentralization of employment opportunities and the possibility of 'profitable' trading down in the housing market from, for example, the expensive areas of Greater London to the much more affordable parts of the countryside.

Employment

As with population, there are considerable similarities between the pattern of employment change across the four rings of the South East and North. Table 4.19 shows that with regard to manufacturing, both Inner London and the Inner Urban North witnessed the greatest percentage decline in employment, followed in turn by Outer London and the Outer Urban North, the Outer Metropolitan Area and Suburban North, and the Outer South East and the Rural and Resort North. Service industry likewise declined in Inner London (and grew only marginally in the Inner Urban North) but experienced its greatest rate of growth in the Suburban North and the Outer Metropolitan Area. Taking employment as a whole, again the Suburban North and the Outer Metropolitan Area were the fastest-growing parts of urban England. From Table 4.19, therefore, 'we can dispel one of the myths about the north. Not only does the north have areas of growth and prosperity, but contrary to popular, but uniformed, assumption they are following a very similar pattern to the South East' (Breheny, Hall and Hart, 1987, p. 25.).

The substantial decline in the number of manufacturing jobs in Inner London (in proportionate terms) was chiefly associated with the closure of the docks and related industries, and the relocation of manufacturing activity to the west of the capital in proximity to the M4 and M3, adjacent rail links and the defence industries of the 'Sunbelt': the manufacturing centre of gravity shifting from east to west across the South East region (Stott, 1988). However, while manufacturing floor-space grew at a rapid rate in the Outer South East, and especially in the west, development was insufficient to offset a net decline in manufacturing employment in this part of the South East.

Keeble (1987) suggested that the urban–rural shift in employment could be explained by any one of the following theories. First, the *constrained location theory* postulated that since site dimensions affect both *in-situ* changes (expansions and contraction) and unit turnovers (openings and closures), cramped sites inevitably accounted for a high rate of loss from both contractions and closures in Inner London and to a lesser extent in the Inner Urban North. Manufacturers, moreover, were unable or unwilling to match the rents paid by competing land-users (such as office firms, retailers and warehouse operators) to secure the space necessary for survival in the inner city. It is notable that in periods of relatively fast economic growth (for example, 1967–75) the decentralization of manufacturing was most marked, while in times of stagnation or recession the pace of decentralization was much slower (such as in the period 1976–83) – the degree of competition for urban sites and the attraction of alternative locations being greater at a time of growth than at a time of recession

Table 4.19 Deconcentration of employment in the South East and North of England, 1971–81

	Manufacturing employment			Service employment			Total employment*		
	1971	1981	Change, 1971–81 (%)	1971	1981	Change, 1971–81 (%)	1971	1981	Change, 1971–81 (%)
Inner London	512,225	309,366	−39.6	1,825,127	1,718,726	−5.8	2,457,491	2,115,244	−13.9
Inner Urban North	1,007,665	714,710	−29.1	1,473,038	1,543,159	+4.8	2,683,532	2,426,096	−9.6
Outer London	537,063	361,153	−32.8	861,959	962,000	+11.6	1,479,733	1,408,046	−4.8
Outer Urban North	671,935	478,584	−28.8	584,592	683,426	+16.9	1,423,245	1,315,117	−7.6
Outer Metropolitan Area	539,253	450,849	−16.4	742,022	948,749	+27.9	1,369,337	1,491,863	+9.0
Suburban North	68,480	60,332	−11.9	103,858	133,025	+28.1	202,232	221,092	+9.3
Outer South East	599,047	523,000	−12.7	1,080,871	1,311,308	+21.3	1,836,118	1,983,445	+8.0
Rural and Resort North	67,304	64,429	−4.3	171,024	196,431	+14.9	276,144	296,566	+7.4

Note
* Includes employment in primary, manufacturing, construction and service industries.

(*Source:* Breheny, Hall and Hart, 1987.)

Table 4.20 House-price/income ratios and house-price changes in selected regions, 1979–81

1979			1980			1981		
	House-price/incomes ratio	Increase in house prices, %		House-price/incomes ratio	Increase in house prices, %		House-price/incomes ratio	Increase in house prices, %
Greater London	3.41	33.8	*South West*	3.06	24.8	*Yorkshire & Humberside*	2.30	12.2
South East (excl. Grt London)	3.46	33.0	*Greater London*	2.96	22.0	*Scotland*	2.44	9.4
South West	3.73	31.7	*Yorkshire & Humberside*	2.34	22.5	*North*	2.16	6.9
Yorkshire & Humberside	2.55	25.3	South East (excl. Grt London)	3.14	21.0	South East (excl. Grt London)	3.04	4.5
North	2.57	22.3	North	2.25	18.2	*South West*	2.88	3.2
Scotland	2.70	19.6	Scotland	2.51	14.7	*Greater London*	2.88	3.1
UK	3.06	29.3		2.78	21.1		2.65	5.5
Average mortgage interest rate, %	11.94			14.92			14.01	

(*Source:* Building Societies Association, *BSA Bulletin*.)

(see Dennis, 1981; Fothergill and Gudgin, 1982; Fothergill, Kitson and Monk, 1985).

Second, the *production cost explanation* posited that higher operating costs in urban areas (directly – the higher cost of land and labour, and indirectly – the cost of congestion and inconvenience) have adverse effects on profitability, investment, competition and employment particularly *vis-à-vis* small firms. Such firms would need to decentralize or risk closure (see Lever, 1982; Moore, Rhodes and Tyler, 1984; Wood, 1987).

Third, the *capital restructuring approach* suggests that in order to apply new techniques and to counter competition, large multi-plant and often multinational corporations shift their operations from cities to rural areas to exploit less skilled, less unionized and less costly labour (see Massey and Meegan, 1978).

Of these three explanations, Keeble (1987) considered the first to be the most credible, while the second approach was still subject to research[2]. The third explanation is sometimes criticized on the grounds that spatial differences in union strength and labour costs are too small to be significant attractions *per se*.

Fothergill and Gudgin (1982) and Fothergill, Kitson and Monk (1985) suggest two further reasons why manufacturing has very largely decentralized. First, the out-migration of population (especially the professional, managerial and skilled-manual classes) has been an important factor in attracting decentralized employment; and, second, planning policies have played a vital part in the decentralization of employment, particularly from London. Through the medium of new towns, expanded towns, industrial-development certificates and office-development permits, 'the coordinated decentralization of population and employment from London [became] ... a central element in strategic planning for the South East in the first thirty years or so after the war' (Buck *et al.*, 1986, p. 53). However, most decentralization has been independent of public policy and almost certainly would have occurred in the complete absence of regional planning.

'Push' and 'pull' effects

For over a hundred years there has been an outward shift of both population and employment from London, and since the inter-war years Britain's other major cities have experienced varying degrees of decentralization. From the 1950s, however, London led the way in decentralizing in absolute terms (Hall *et al.*, 1973).

From an alternative perspective, Hall (1987a) offered a further explanation of the urban–rural shift. He suggested that decentralization occurs as a result of 'push' and 'pull' effects. In London, limited space and an often unsatisfactory environment have pushed out a proportion of the relatively affluent to beyond the capital's suburbs, while in large cities, in general, the structural decline of industry pushes out firms in search of preferable operational opportunities elsewhere. Pull effects exert an important influence when easier access to motorways, a more stable and less militant labour-force, possibly lower wages and a more pleasing work-inducing environment attract manufacturing firms from the cities. With regard to service employment, lower rents and salaries have attracted firms wishing to decentralize their large-scale clerical-processing activities, and decentralized locations have also attracted producer services (banking, insurance, accountancy and legal services) as well as public-sector employment (in, for example, local government, further and higher education and health care). These jobs in turn have created a demand for a wide range of consumer services such as retailing, schools, entertainment, building and estate agencies. Small rural or coastal towns also attract the retired - and subsequently a range of relevant service employment. Thus, whereas until the 1950s there was a stark dichotomy between the economic potential and amenities of the urban and the rural areas of Britain, both north and south, this gulf is no longer evident.

Housing: the problems of the south

Although undoubtedly there has recently been a stark north-south divide in house prices (see Chapter 2), housing analysts and commentators often overlook long-term trends in house-price variations, intra-regional disparities in values, the land question in the South East and problems of housing in London.

A consideration of long-term trends reveals that interregional variations in house prices fluctuate considerably. Figure 4.6, for example, shows that throughout the period 1969-88, the widest gaps between the highest house-price region (Greater London) and the lowest-price regions (the North and Northern Ireland) occurred in 1972-3, 1979-80 and 1987, whereas in the intervening years there was a significant narrowing of the gap as average house prices converged. Within this apparent seven-year cycle, convergence after 1972-3 and, again (to a much lesser extent), after 1979-80, was attributable to an above-average increase in house prices in the north after the house-price boom had collapsed in the four southern regions (Hamnett, 1988b).

Clearly, in the north comparatively low house-price/incomes ratios ensured that demand and house prices decreased fairly gradually after mortgage interest rates were increased in 1973-6 and 1979-81, whereas in the south (and in Greater London in particular) high house-price/incomes ratios brought about a sudden reduction in demand and house prices during the same periods. Table 4.20 shows that when mortgage interest rates were relatively low (in 1979), the highest increase in house prices occurred in Greater London - despite the house-price/incomes ratio being comparatively high in the capital. However, after mortgage interest rates had stayed high for two years, Yorkshire & Humberside, Scotland and the North showed the greatest increase in house prices: very probably because of the low house-price/incomes ratios in these regions cushioned much of the impact of higher interest charges. Despite opportunities for capital accumulation, house-buyers in the South East were eventually unwilling to pay higher and higher prices for properties - which, price for price, were smaller and of poorer quality than their northern equivalent - and at the same time they were reluctant to cut back dramatically on their general consumption in order to pay off more expensive mortgages. In general, house-buyers in the north, in contrast, did not feel overstretched when interest rates rose

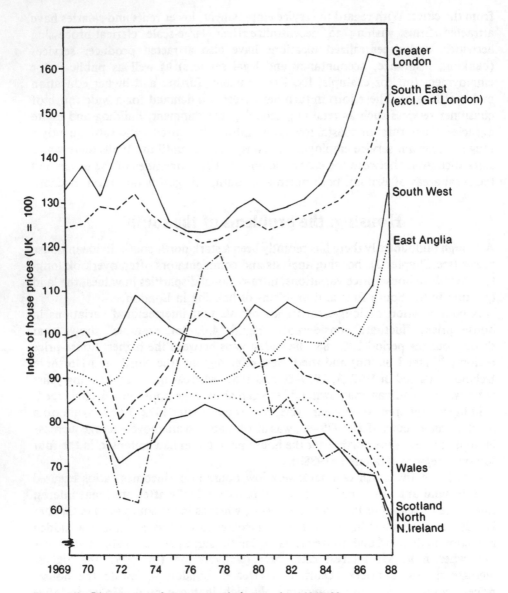

Figure 4.6   Divergence and convergence in house prices, 1969-88
          (Source: Building Societies Association)

and they were thus more able to maintain both their demand for housing and
other forms of consumption.

The escalation of mortgage interest rates (from 9.5 to 13.5 per cent, April 1988
to January 1989) signalled a further convergence in house prices. Higher interest
rates had a particularly severe effect on owner-occupiers (and particularly first-
time buyers) in the South East since they were the largest borrowers, while, by
comparison, mortgagors in the north were little affected. Whereas house prices in
Greater London rose by 'only' 15.6 per cent between the first quarter of 1988 and

the first quarter of 1989, in the north house prices continued to surge ahead – for example, by as much as 54.4 per cent in the East Midlands, 52.6 per cent in Yorkshire & Humberside and 49.5 per cent in Wales. In early 1989 it was thus evident that whereas the house-price boom was over for the time being in the south, it continued apace in much of the north (Figure 4.7) – bringing about a further convergence of regional house prices.

Although reference is rarely made to interregional disparities in the long-term house-price cycle, price variations within regions receive possibly even less comment. Referring to 1985 data, Champion, Green and Owen (1988) showed, however, that at the level of the local labour-market area the greatest variation in intra-regional mean house prices (£33,000 in the South East) was substantially greater than the widest interregional variation in prices at the level of the standard region (£19,809).

Intra-regional variations are also apparent when house prices are examined at the level of the borough or district. Table 4.21 reveals that in 1988 there was a difference of £71,800 between the mean price of housing in the most and least expensive areas of the South East (Haringey and South Wight) compared to a difference of 'only' £59,706 between the highest- and lowest-price standard region (the South East and North). As Champion, Green and Owen had shown, there were significant price disparities within all regions, although none were as wide as in the South East. Within regions, as between regions, house-price disparities impede household mobility. It is financially difficult for households to migrate from low- to high-price areas, from country towns to the major cities, but whereas migration from high- to low-price areas has for long been common, it is normally undertaken with the knowledge that a return move may prove impossible. Clearly within most regions, but particularly in the South East, there is a trade-off between housing costs and travel costs – with the South East experiencing an increasingly severe problem of rail and road commuting, the journey to work being the dominant cause of traffic congestion.

Disparities in land values are, of course, very evident. Table 4.22 reveals that the price of plots as a proportion of the average price of new dwellings varied from 53 per cent in Greater London to 12 per cent in Wales in 1987. Regional land prices were not only dramatically higher (both relatively and absolutely) than in 1979 (Table 4.22), but also in most regions (and particularly in Greater London) increased much more quickly than either house prices or the retail price index. While it is conventional wisdom to argue that house prices determine plot prices (see Drewett, 1973; Harloe, Issacharoff and Minns, 1974; Bassett and Short, 1980; Merrett with Gray, 1982; Ball, 1983), there are circumstances when it could be claimed that land costs influence house values. Research undertaken by Evans (1987) for the House Builders Federation suggested that a restricted supply of land, particularly in the South East, resulted in both land prices and house prices being higher than they would be otherwise; that planners – by restricting the supply of building land – were forcing up house prices; and that if agricultural land was converted into residential use it would significantly bring down the cost of new housing.

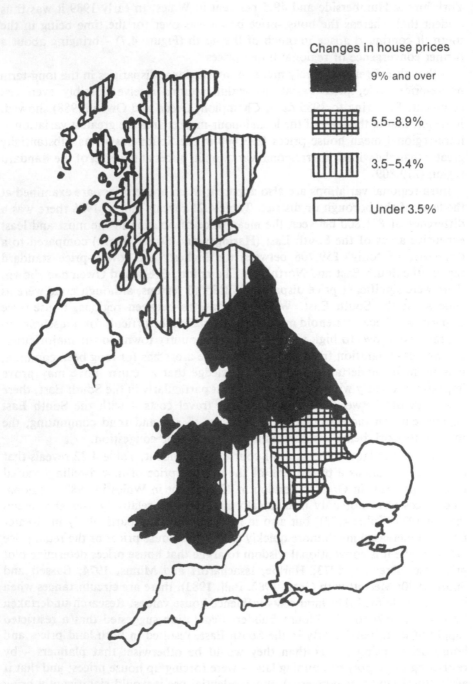

Changes in house prices

9% and over

5.5–8.9%

3.5–5.4%

Under 3.5%

Figure 4.7   Changes in house prices, UK, first quarter 1989 (Source: Nationwide Anglia Building Society)

Table 4.21 Interregional and intra-regional disparities in house prices, Great Britain, 1988

Region	Mean price (£)	Highest-price district or borough	Mean price (£)	Lowest-price district	Mean price (£)	Difference (%)
South East	90,099	Haringey	113,750	South Wight	41,906	271
East Anglia	51,220	South Cambridgeshire	64,560	Great Yarmouth	39,791	247
West Midlands	35,995	Solihull	56,990	Staffordshire Moorlands	23,175	246
East Midlands	32,855	South Northamptonshire	50,288	Newark	23,734	212
Wales	29,946	Cynon	40,454	Llanelli	20,533	197
Scotland	38,281	Aberdeen	50,437	Angus	26,955	187
Yorkshire & Humberside	31,175	Harrogate	44,168	Barnsley	24,101	183
South West	48,232	Christchurch	62,425	North Cornwall	35,100	178
North West	31,425	Macclesfield	43,394	Halton	25,128	173
North	30,393	Gateshead	36,850	Easington	22,708	162
Maximum difference in interregional house prices, %						301

Notes
1. To ensure that comparisons are made on a 'like-with-like' basis, the above prices refer exclusively to three-bedroomed semi-detached houses.
2. Year ended March 1988.

(*Source:* Nationwide Anglia Building Society.)

Table 4.22   Cost of land and new-dwelling prices, England and Wales, 1979–87

	Average price per plot, £		Price of new dwellings, £		Plot price as a percentage of new-dwelling price	
	1979	1987	1979	1987	1979	1987
Greater London	6,867	32,224	27,913	60,800	24.6	53.0
Rest of South East	4,138	23,293	25,800	65,739	16.0	35.4
South West	3,869	13,370	20,504	52,293	18.9	25.6
West Midlands	3,449	12,860	20,724	48,654	16.6	26.4
East Anglia	2,206	8,678	18,863	50,110	11.7	17.3
North	2,853	7,905	17,951	41,216	15.9	19.2
East Midlands	1,534	7,645	17,920	42,619	8.6	17.9
North West	3,222	7,210	19,595	41,754	16.4	17.3
Yorkshire & Humberside	1,658	4,725	18,426	39,264	9.0	12.0
Wales	1,686	4,557	18,316	38,122	9.2	12.0

(*Source*: Department of the Environment, *Housing and Construction Statistics*.)

The conflict between house-building and conservation was particularly apparent in the South East. While house-builders predicted that there was a need for an extra 880,000 dwellings in the region in the period 1981–91, the South Eastern Regional Planning Conference (SERPLAN) allocated land for only 600,000 new dwellings (*Economist*, 1985). Similarly, whereas Tyler and Rhodes (1986) forecast that an additional 770,000 new homes would be required in the region during the years 1985 to 1995, SERPLAN predicted that only 550,000 extra dwellings would be needed. For the decade 1991–2001, the forecasts of house-builders and planners diverged even more substantially – the counties initially failing to allocate sufficient land to enable the construction of the number of houses that they themselves forecast would be needed (Hall, 1985). It is very unlikely, however, that even the eventual official house-building target of 570,000 dwellings (set by the Secretary of State for the Environment in 1988) will be adequate to satisfy demand. Restrictions on the supply of building land has inevitably created a chronic housing shortage in the South East, and in the late 1980s caused a dramatic hike in the rate of house-price inflation in the region (Lock, 1988). Early in 1988, average house prices in the South East were soaring at an annual rate of over 20 per cent (greatly ahead of the increase in retail prices) and by the end of the year had reached £92,160 in Greater London and £95,320 in the Outer Metropolitan Area (Nationwide Anglia Building Society, 1989). This not only made housing in the South East unaffordable to most northerners, but also put it out of reach to many potential first-time buyers in the region itself (TCPA, 1987). According to the London Research Centre (LRC) (1988, p. 76), 'only a quarter of potential households wishing to buy in London were in a financial position to purchase even the very cheapest properties available in the capital'.

The housing shortage in Greater London was exacerbated by the drift of population to the capital in search of work, young people being driven or evicted from their homes, family break-up, unemployment among vulnerable groups, escalating rent levels and the selling-off of private rented housing (Eddison, 1988). According to the LRC (1988, p. 76), the private rented sector declined by approximately 17,000 dwellings per annum in the mid-1980s, while council-house sales in the capital further reduced the availability of rented housing. Although only 9.6 per cent of Greater London's council stock was sold off in the period 1979-85, a total of 86,954 rented dwellings were consequently lost to this sector – more than in any region except the rest of the South East (Dunn, Forrest and Murie, 1987). Shortages were also attributable to the poor condition of the housing stock. *The Greater London House Condition Survey* (Greater London Council, 1985b) found that 22 per cent of the capital's housing was either unfit for human habitation or lacking in basic amenities or in serious disrepair, and in parts of inner London (such as Hackney), the proportion of unfit housing was almost double the average for London as a whole.

Table 4.23   Homelessness, Great Britain, 1987

	Total households accepted	Acceptances per thousand households
	('000)	
South East	47,230	7.1
(Greater London)	(29,710)	(11.0)
(Remainder of South East)	(17,520)	(4.0)
North West	17,770	7.4
North	8,360	7.1
West Midlands	12,000	6.2
Scotland	11,290	5.8
East Midlands	7,660	5.1
Yorkshire & Humberside	9,220	4.9
Wales	4,570	4.4
South West	7,210	4.1
East Anglia	2,990	4.0
Great Britain	128,200	6.1

(*Sources*: Department of the Environment; Scottish Development Department; Welsh Office.)

Different indicators have been used to show the degree of housing need in the capital. First, there were a quarter-of-a-million concealed households in London in 1986-7 living as part of someone else's household but who would have preferred separate housing. Within this group there were up to 59,000 lone-parent and couple-potential households requiring a home of their own. Second, there were 12,500 squatters in the capital in 1986-7 – about half over the age of 25; and, third, in the same year, there were 127,000 households registered on council waiting-lists and 20,000 households on housing-association lists (LRC, 1988, p. 76). Finally, and most seriously, whereas Greater London contained only one-eighth of the population of Great Britain in the mid-1980s, the capital had more than one-quarter of the country's homeless (Table 4.23), and the extent of homelessness had increased dramatically since the early 1980s. In 1987, 29,710 families were accepted as homeless by the London boroughs (an increase of 70 per cent over 1980-1), but countless other people were homeless but not accepted by the boroughs under the provisions of the Housing (Homeless Persons) Act 1977. A disturbing new trend was the big increase in the number of young homeless people, rising to 64,000 in the capital in 1987, according to the London Housing Unit.

Apart from the failure of government to satisfy the housing needs of the disadvantaged, planning policies in the 1980s largely ignored the fundamental aspirations of many ordinary households in the South East (Lock, 1989). In protecting the Green Belt and in attempting to revitalize London, planning authorities seemed to assume (wrongly in the view of Lock) that newly formed households in the region preferred to live in (rehabilitated/converted) flats in inner London than in houses in the shire counties; that low-income inner-city households (with children) no longer wished to move out to the suburbs or beyond (the new-town and expanded-town programmes being progressively wound down since the mid-1970s); and that most households found high-density housing (for example, flat conversions and infill developments) more suitable for their needs than lower-density housing and more green space.

It was ironic that, although the Department of the Environment and the county councils of the South East strongly defended the Green Belt against urban/suburban encroachment, they became embroiled in the planning of a number of massive infrastructure developments in the region – most notably the M25, M11 and M40, Stansted airport and the Channel tunnel. At both a central- and local-government level, however, there was a failure to relate infrastructure investment to potential housing, industrial and commercial development. Instead, the government regarded the resulting (or impending) overheating of the economy of the South East as a problem rather than an opportunity for positive regional planning. Retailing and industrial developments, for example, were severely constrained close to the M25, there was an absence of any planning strategy *vis-à-vis* the M11 and M40, the Essex and Hertfordshire structure plans ignored the probable development spin-offs from the expansion of Stansted, and SERPLAN failed to consider the likely effects of the Channel tunnel upon the economy of region as a whole (see Lock, 1989). Clearly, the housing crisis in Greater London (and particularly in inner London) was just as much attributable to the weaknesses of planning policy in the South East region as to the imperfections of housing policy in the capital itself.

The health divide: alternative perspectives

While there are striking interregional differences in health (death rates in the 1970s to 1980s being highest in the north and lowest in the south – see Chapter 2), the spatial distribution of specific diseases do not always conform with this disparity. With regard to breast cancer, for example, Gardner *et al.* (1983, 1984) detected a distinct concentration of this disease in the south. Since the risk of breast cancer is greater among women who have had a small rather than a large number of children, and as birth rates are lower within social classes I and II, the greater concentration of these classes in the south might account for a greater incidence of this disease in the southern regions of Britain. Gardner *et al.* also found a greater concentration of stomach cancer among women in the south (for no apparent reason) and higher rates of melanoma (among both men and women) due to more intermittent intensive sunbathing. Concentrations of melanoma in the south are almost certainly attributable to climatic factors and affluence (more holidays abroad being taken by southerners than northerners). Male suicide rates are particularly high in London, and rates are high among the elderly of both genders in Cornwall and Devon, where the incidence of suicide in the over-75 age group is more than three times greater than among the 15–25 age group.

The spatial distribution of disease reflects to a significant extent past environmental and life-style determinants. Present conditions will influence future patterns of disease and mortality and might conceivably reverse the general north-south inequality of health. While Hancock and Dahl (1986) argue that a healthy city (and by implication a region) continually creates and improves its physical and social environment enabling people to develop themselves to their maximum potential, Ashton suggests that, as we move towards 2000, 'this prospect seems to be slipping away from those living in the south-east of England' (Ashton, 1988, p. 278).

More specifically in the south east there is gross overcrowding, barely affordable housing, long travel times and expensive journeys to work and to the countryside, difficult access to the capital's culture (in terms of transport) compared to counterparts in the provinces, and often an absence of kinship networks to support those newly arrived in the region. In general, there is a 'stressful environment with a poor quality of life which must affect health in the future' (Ashton, 1988, p. 278). In contrast, claims Ashton, there are distinct benefits (at least to the managerial and professional classes) from living in the north. For example, there are higher-quality working conditions, cheaper housing, shorter travel times, easier access to cultural and recreational facilities, and relatively empty roads and lack of traffic jams. Many of the causes of ill health in the north were legacies of the Industrial Revolution and have been substantially eliminated, while the emergence of a more healthy environment is becoming a key attraction to both capital and labour.

Apart from interregional disparities in health there are marked intra-regional inequalities. Howe (1986) has shown that in the case of acute myocardial infarction (coronary heart disease) among males in Greater London, standard mortality ratios (SMRs) ranged from 122 in Dagenham and 117 in Newham to 56 in Croydon and 51 in Sutton; and in respect of lung cancer, SMRs ranged from 165 in Southwark and 137 in Islington to 61 in Barnet and 58 in Kingston upon Thames.

Similar inequalities exist in other urban areas. Carstairs (1981) detected greater mortality and morbidity in areas of greater deprivation in both Edinburgh and Glasgow; and the Greater Glasgow Health Board reported in 1984 that in several communities within the city death rates were among the worst in the world, whereas in many other communities within its area death rates were as low or even lower than in the healthiest cities in Europe. In Sheffield, Thunhurst (1985) found that there were large differences in mortality between the most affluent and most deprived wards; and in considering 36 geographical clusters of wards in England and Wales, Fox, Jones and Goldblatt (1984) detected patterns of high mortality in low-status clusters and low mortality in high-status clusters.

Whitehead (1988) suggested that such intra-regional or intra-urban disparities demonstrate that the so-called north-south health divide is far too simple. The healthiest areas in parts of the north, she argued, compare well with the healthiest areas in parts of the south; conversely, in both the north and the south, poor health is associated with the various indicators of material and social deprivation – similarities that are often masked when analyses are carried out at the level of the county or regional health authority.

Table 4.24   Projected population increase, 1986–2001, and hospital waiting-lists for non-casualty treatment, 1989

	Projected population increase, 1986–2001	Health authority	Patients waiting more than 12 months*
	('000)		(%)
South East	1,079	West Surrey and North-East Hants	55
		West Lambeth	53
		Basildon and Thurrock	45
		Barking, Havering and Brentwood	44
		Tower Hamlets	42
		Islington	42
		Hampstead	41
South West	440	Frenchay, Bristol	51
		Salisbury	40
West Midlands	128	South Birmingham	44
		North Warwickshire	41

Note
* In 1989.

(*Source*: Office of Population Censuses and Surveys; Ferriman, 1989.)

By the end of this century, regional differences in morbidity and mortality rates might well be attributable, in part, to variations in hospital waiting-lists. Figures issued by the Department of Health in January 1989 showed that waiting-lists for all (non-casuality) treatment were longest in the south, and were particularly lengthy in West Surrey, North East Hampshire, West Lambeth and in Bristol (Table 4.24). It was clear that, in parts of the south, population was increasing at a substantially faster rate than the allocation of resources to the National Health Service: a disparity that is likely to worsen if, for example, the population of the South East increases by a million people over the period 1986 to the end of the century, as is projected (Ferriman, 1989).

The prosperity divide: a reinterpretation

Although recent analyses have shown that in aggregate there is a distinct north-south divide in prosperity (Champion and Green, 1985, 1988; Mintel, 1988), this finding must be strongly qualified to avoid misinterpretation.

Whereas research shows that most local labour-market areas (LLMAs) in the south are more prosperous than those in the north, Champion and Green (1988) acknowledge that the south is not uniformly prosperous nor the north uniformly poor. Although the ten worst-performing LLMAs in the south, for example – according to Champion and Green's amalgamated index – were ranked within the range 151–261 (within a total of 280 LLMAs in Britain), the ten-best-performing LLMAs in the north were ranked within an appreciably higher range, 32–68 (Table 4.25).

Table 4.25 Worst-performing local labour-market areas in the south, and best-performing market areas in the north, 1981–7

Bottom-10 LLMAs in the south			Top-10 LLMAs in the north		
Rank	LLMA	Score	Rank	LLMA	Score
151	Margate & Ramsgate	0.429	32	Stratford-upon-Avon	0.618
172	Dereham	0.412	42	Northampton	0.577
174	Yarmouth	0.412	46	Corby	0.570
186	St Austell	0.402	49	Macclesfield	0.568
187	Launceston	0.401	51	Northallerton	0.565
189	Woodbridge	0.401	56	Kettering	0.554
195	Bideford	0.396	59	Buxton	0.548
213	Lowestoft	0.377	60	Kendal	0.540
241	Deal	0.345	61	Hinckley	0.539
261	Redruth & Camborne	0.321	68	Wellingborough	0.529

(*Source*: Champion and Green, 1988.)

Analyses at the level of the LLMA also show that there are very substantial intra-regional variations in prosperity. For example, in the South East the highest-scoring LLMA, Milton Keynes, is the best performing area in the whole of Great Britain, while the worst-performing LLMA in the region, Deal, is ranked 241. Even in the region with the least intra-regional disparity, Wales, there is a difference in rank of as much as 177 between Aberystwyth (ranked 103) and Holyhead (ranked 280) (Table 4.26).

Possibly the strongest reason for arguing that analyses at the level of the LLMA or the region are somewhat over simple is that they fail to take account of the high incidence of deprivation in parts of Greater London – the capital as a whole being regarded as just one LLMA, and as such being seen to perform fairly satisfactorily – ranking 38th out of 280 in Champion and Green's 1988 amalgamated index. Yet London's wealth has manifestly failed to reach many sections of the capital's population. The Association of London Authorities (ALA) (1988) pointed out that the Z score, which measures multiple deprivation

Table 4.26 Intra-regional comparisons of best- and worst-performing local labour-market areas, Great Britain, 1981–7

Region	Highest-scoring LLMA	Rank	Lowest-scoring LLMA	Rank	Differences in rank
South East	Milton Keynes	1	Deal	241	240
West Midlands	Stratford-upon-Avon	2	Smethwick	269	237
East Midlands	Northampton	42	Mansfield	272	230
Yorkshire & Humberside	Northallerton	51	Mexborough	279	228
North West	Macclesfield	47	St Helens	273	226
South West	Salisbury	37	Redruth & Camborne	261	224
North	Kendal	60	Consett	268	208
East Anglia	Cambridge	6	Lowestoft	213	207
Scotland	Hawick	89	Stranraer	275	186
Wales	Aberystwyth	103	Holyhead	280	177
Great Britain	Milton Keynes	1	Holyhead	280	279

(*Source*: Champion and Green, 1988.)

from data derived from the 1981 *Census of Population* (regarding unemployment, poor housing and overcrowding, poor health and the number of single-parent families), showed that 16 of the 25 most deprived areas in England and Wales were located in London (Table 4.27). The ALA also indicated that in recent years there had been a divergence in the living standards of the wealthy and the poor in the capital. For example, the real incomes of the poorest sections of London's population had diminished since 1979, inequality of income had become more marked and polarization had become more pronounced than in the rest of the UK. These findings were supported by the *New Earnings Survey* (Department of Employment, 1985), which showed that, in the period 1979–85, the average earnings of the best-paid 10 per cent of full-time male workers in London had increased from 167 to 183 per cent of median earnings, whereas the average earnings of the lowest-paid 10 per cent had decreased from 64 to 60 per cent of the median.

The ALA (1986) reported that by the mid-1980s a quarter of the capital's full-time workers had earnings below the Low Pay Unit's poverty line, and Lock (1988) – referring to chronic inner-city poverty in London – claimed that two-thirds of women and one-sixth of men in London earned less than £100 per week in full-time jobs in 1987; that the real income of the poorest 25 per cent of

Table 4.27   Areas of deprivation in England and Wales as measured by the Z-score index of multiple deprivation

Rank	London boroughs		Rank	'Northern' district authorities	
1	Hackney	6.69			
2	Newham	5.84			
3	Tower Hamlets	5.53			
4	Lambeth	5.52			
5	Hammersmith & Fulham	4.90			
6	Haringey	4.86			
7	Islington	4.80			
8	Brent	4.62			
9	Wandsworth	4.50			
10	Southwark	4.40			
			11	Manchester	4.19
12	Camden	4.05			
			13	Leicester	4.02
			14	Wolverhampton	3.77
			15	Birmingham	3.51
16	Lewisham	3.46			
17	Kensington & Chelsea	3.34			
			18	Coventry	3.17
			19	Sandwell	3.15
			20	Nottingham	3.14
21	Waltham Forest	3.09			
			22	Blackburn	3.08
23	Westminster	2.97			
24	Croydon	2.90			
			25	Knowsley	2.67

(*Source*: Association of London Authorities, 1988.)

Londoners fell by 4.3 per cent (1971–81); and that one in six London households depended almost entirely on welfare benefits in the mid-1980s – at least one-third up on 1979. Conditions of work also deteriorated for many people in the capital. The Greater London Council (1985a) was concerned that, since 1979, the amount of shift work, casual employment, part-time work and sweating had increased, particularly in partly skilled and unskilled manual occupations – in addition to a worsening in pay in hotel work, cleaning, food processing and branches of the retail trade.

In the inner-London boroughs in the late 1980s, deprivation was acute. In Newham, for example (the second-poorest local-authority district in England and Wales), unemployment in 1987 reached 20 per cent, a large proportion of its population was dependent on benefit payments, over 16,000 private dwellings were in urgent need of repair, while the council-house stock was being diminished by a programme of demolition. The same was generally true of many other inner-London boroughs – even the Royal Borough of Kensington and Chelsea had an unemployment rate of 7.2 per cent in 1987, twice that of Cheadle in Greater Manchester (Wilsher and Cassidy, 1987).

Clearly, the south is not entirely a 'land of milk and honey'. Affluence is not much in evidence in the many depressed and decaying industrial, market or resort towns in the South East, South West and East Anglia, nor in the appallingly deprived areas of London. The standard of living of even the relatively well-paid is lower in the south than in the north. The Reward Group (1989) (using a *quality of life index* to compare salaries and the cost of living in different regions), revealed that middle managers were badly off in the South East and Greater London and comparatively well off in, for example, Scotland and the North. Although in London a middle manager's salary is 17 per cent above the national average, the cost of living is 39.9 per cent higher – the middle manager having a quality-of-life index of − 22.9 (Table 4.28). Clearly, in the late 1980s, middle managers in the south would dramatically improve their material standard of living, at least in the short term, if they migrated northwards.

Although regional aggregate figures (examined in Chapter 2) very largely present the north as if it were a homogeneously depressed whole, lack of

Table 4.28   Regional standard of living: middle management, 1989

	Quality-of-life index
Scotland	+ 19.7
North	+ 14.2
Yorkshire & Humberside	+ 10.4
North West	+ 5.9
East Midlands	− 3.6
West Midlands	− 6.7
South West	− 9.4
East Anglia	− 9.6
Greater London	− 22.9
South East	− 23.0

(*Source*: Reward Group, 1989. Diamond Way, Stone, Staffs.)

economic uniformity is no less evident than in the south. Figures are often too crude, too simple and in consequence misleading. Disaggregated data, however, show that in terms of performance and opportunity, many towns in the north compare very favourably with the most prosperous urban areas of the south.

As a preliminary to identifying those towns in the north that showed the greatest potential for growth, Breheny, Hall and Hart (1987) presented a fine-grain analysis of material and qualitative living standards in the North West, the Northern region and Yorkshire & Humberside. Initially they pointed out that unemployment rates in many northern travel-to-work areas were at or below rates for the south - for example, at 8 per cent in Matlock, 8.2 per cent in Ripon or 8.3 per cent in Macclesfield compared to over 15 per cent in many inner-London boroughs or more than 20 per cent in Thanet (Kent) in 1986; and that the diversion of maritime trade from North and South America and the Commonwealth to Europe was creating a west-east divide (with depressed ports and hinterlands in the west and growing opportunities in the east).

Objective indicators were then used in an attempt to measure the character of life in the north. With regard to physical environment, housing, transport, education, health and cultural provisions, Breheny, Hall and Hart (p. 18) claimed that 'in the last quarter century, the north has been ... transformed beyond recognition [and] ... in many important respects the north now offers its residents a quality of life far superior to that in and around the metropolis'. Housing, for example, is generally cheaper in the north than in most of the south, enabling owner-occupiers in the north to benefit from higher disposable incomes after meeting housing costs. Clearly the conclusion drawn from 'the evidence on incomes and house prices [alone] is that the standard of living of those in work in the north of England is likely to be higher, on the basis of income received pound for pound, than in the south' (*ibid.* p. 29-30).

Whereas Champion and Green (1985, 1988) constructed indices to measure material prosperity (a reflection of past economic performance), Breheny, Hall and Hart considered that there was a need for a broader index to identify those areas in the north that showed the greatest potential for growth. Using official figures from the 1981 *Census* and subsequent government surveys, Breheny, Hall and Hart therefore produced an index based on an average of nine separate socio-economic indicators. This showed that out of the 92 districts in the North West, North and Yorkshire & Humberside, ten (namely Clitheroe, Harrogate, Morpeth, Knutsford, Kendal, Lytham St Annes, Thirsk, Hexham, Beverley and Congleton) were found to be 10-20 per cent more desirable than the average local-authority area in Great Britain as a whole. A more detailed analysis of these towns was then undertaken (using a more local set of indicators) to produce both a *quality index* and a *service index* (Table 4.29). In conclusion, Breheny, Hall and Hart (p. 38) showed that the ten towns - the 'Northern Lights', as they were termed - offered 'a quality of life at least equal to - and in important respects, such as access to the countryside, superior to anything enjoyed by their equivalents in the south. And they do so at a fraction of the cost of their [southern] counterparts'.

By the late 1980s the media began to examine whether or not the north-south

Table 4.29 The prosperity, quality and service indices of the 'Northern Lights'

	Cl.	Ha.	M	Kn.	Ke.	LSA	T	He.	B	Co.	England
Prosperity Index											
Social class 1 & 2	37.00	33.50	38.30	41.10	29.20	32.10	34.20	31.70	36.70	35.50	23.30
% higher ed. qual.	19.90	17.30	22.60	22.00	17.20	17.70	16.00	16.20	19.50	17.20	13.00
% male unemployment	4.90	6.10	6.10	7.20	6.00	7.00	7.40	7.10	6.90	10.10	11.60
% youth unemployment	10.00	8.80	10.30	15.00	9.60	12.60	12.80	11.00	15.20	15.10	19.30
% 1 + car	73.40	68.60	70.80	72.90	71.30	67.80	75.50	65.60	72.90	75.50	60.50
% owner-occupation	75.60	68.10	54.50	68.60	63.90	74.50	61.30	53.10	72.40	72.90	55.90
Persons/room	0.51	0.49	0.53	0.50	0.48	0.48	0.50	0.52	0.51	0.52	0.55
5 + O-levels, No A-level	16.60	11.20	13.70	11.60	10.60	16.60	11.20	13.70	10.00	11.60	10.80
1 + A-levels	9.70	20.30	17.20	15.10	12.60	9.70	20.30	17.20	13.80	15.10	13.90
Average percentage of England comparable	148	143	143	137	131	131	130	129	128	123	100
Quality Index											
Restaurant (*Good-Food Guide*)	5m	Yes	Yes	Yes	—	Yes	12m	6m	6m	Yes	
Hotel (*Michelin*)	3m	Yes	Yes	Yes	—	Yes	Yes	Yes	Yes	No	
Antique shops	1	58	1	8	—	3	No	15	4	1	
Golf courses	Yes	Yes	Yes	No	—	Yes	Yes	Yes	Yes		
National park/AONB	2m	10m	5m	13m	—	18m	5m	5m	20m	12m	
Service Index											
Service-centre grade	4B	3B	4B	4B	4B	4B	4B	4A	4A	4B	
Distance major centre	8m(3A)	Nil	12m(2B)	16m(2A)	21m(3A)	6m(3A)	23m(3A)	12m(2B)	8m(2C)	12m(3A)	
Distance to motorway (miles)	8m	6m	3m	2m	6m	5m	1m	1m	12m	7m	
Commuter rail	No	Yes	Yes	Yes	Yes	Yes	Yes	Yes	Yes	Yes	

Notes
Cl. Clitheroe;   Ha. Harrogate;   M Morpeth;   Kn. Knutsford;   Ke. Kendal;   LSA Lytham St Annes;   T Thirsk;   He. Hexham;   B Beverley;
Co. Congleton;   m miles;   2A,B,C provincial centres;   3A,B major and intermediate regional centres;   4A,B minor regional centres.

(*Source*: Breheny, Hall and Hart, 1987.)

divide was a reality or a myth. 'Like most such generalisations', argued *The Sunday Times* (1987a), the phrase 'north-south' divide 'obscures more than it illustrates'. For example,

> few towns in the south can match the prosperity of Aberdeen or Edinburgh ... and industrial decay is not much in evidence in Harrogate and its environs, or in the Manchester stockbroker belt of North Cheshire, where the residents from the Home Counties could learn a thing or two about affluence.

Writing in the same newspaper, Wilsher and Cassidy (1987) drew attention to the very visible signs of affluence in the North region; to the proliferation of large-scale retail centres in the north east (for example, the Tees Bay Retail Park at Hartlepool and the Metrocentre at Gateshead – the most extensive shopping and leisure centre in Europe); to the attraction of the region to foreign firms (for example, Nissan at Washington); and to the large volume of private investment that was being undertaken in the region (£600 million in 1987).

While the above evidence suggests that both the quantitative and qualitative attributes of parts of the north are comparable to those of much of the south, studies by Rogerson, Findlay and Morris (1988) and Rogerson *et al.* (1989) stress that social rather than economic features are the most important elements in the quality of life for the majority of people in Britain. In their initial study, Rogerson, Findlay and Morris (1988) first ranked the largest city regions in Britain (those with populations in excess of 250,000 in 1981) in order of their social, economic and physical characteristics as indicated by data derived from official sources. Second, non-unitary weightings were then attached to each of the characteristics on the basis of their perceived importance (as determined by opinion surveys), and the magnitude of each characteristic is then multiplied by its weighting. Third, the final ranking was achieved by summing the scores of all the characteristics. Table 4.30 not only shows that Edinburgh and Aberdeen are first and second in the quality-of-life ranking, but out of the top-10 city regions eight are situated in the north. It is notable that the opinion surveys used in this analysis showed that people (if given the choice) would choose to live in a place that had (in descending order of preference) low levels of crime, good health-service provisions, low levels of pollution, a low cost of living, good shopping facilities and racial harmony.

In their second study (and using broadly the same approach), Rogerson *et al.* (1989) ranked 34 'intermediate-sized cities' on the basis of local labour-market areas, and this too showed that the perceived quality of life, in general, was higher in the 'north' than in the 'south'. Although Exeter came out on top, Halifax, York, Dundee and Swansea – all in the 'north' – ranked second to fifth respectively, whereas – in the South East – Guildford ranked 20th, Maidstone 23rd, the Medway towns 28th and Slough 31st.

Although it might seem that there is a growing uniformity between regions (there are the same chain-stores in every town with the same consumer goods on display, and urban individuality is being reduced by the decreased power of local government), a report by Mintel (1988) suggests that the quality of life in the north has been greatly enhanced by the development of regional television, the

Table 4.30   Quality-of-life ranking, Great Britain

Rank	City/town	Rank	City/town
1	Edinburgh	20	Derby
2	Aberdeen	21	Norwich
3	Plymouth	22	Birkenhead
4	Cardiff	23	Blackpool
5	Hamilton–Motherwell	24	Luton
6	Bradford	25	Glasgow
7	Reading	26	Bournemouth
8	Stoke-on-Trent	27	Leeds
9	Middlesbrough	28	Sunderland
10	Sheffield	29	Bolton
11	Oxford	30	Manchester
12	Leicester	31	Liverpool
13	Brighton	32	Nottingham
14	Portsmouth	33	Newcastle
15	Southampton	34	London
16	Southend	35	Wolverhampton
17	Hull	36	Coventry
18	Aldershot–Farnborough	37	Walsall
19	Bristol	38	Birmingham

(*Source*: Rogerson, Findlay and Morris, 1988.)

establishment of polytechnics and new universities, the revival of the arts, the greater acceptance of regional accents and the cleaning up of towns and cities. The virtual absence of commuting problems in the north is an increasingly attractive qualitative attribute when compared to worsening traffic congestion in the South East. It is perhaps thus understandable that, according to the Mintel report, a significantly lower proportion of interviewees in the north would like to move to another part of the country compared to those in the south (for example, 22 per cent in the north east and 30 per cent in Yorkshire, in contrast to 41 per cent in London).

Overall, in terms of both economic performance and the quality of life, it is a very great over-simplification to talk about a north-south divide. There is both considerable deprivation in parts of the south (particularly in London) and very evident prosperity and growth potential in areas of the north. How this affects the electoral fortunes of the major political parties will be examined in the following section of this chapter.

The diversity of electoral support

Although, as Chapter 2 demonstrates, there was a clear north-south divide in support for the Conservative and Labour Parties throughout the 1980s (the Conservatives dominating the south and Labour being strongest in the north), the pattern of disparity is more complex when examined at a disaggregated or localized level.

In the north in 1987, the Conservatives were able to win a substantial number

Table 4.31   Unemployment and political support: Conservative- and Labour-won seats in northern regions and Greater London, May 1988

	Unemployment (%)		Unemployment (%)

Northern Conservative-won seats

North
Barrow & Furness	7.7
Darlington	12.1
Langbaurgh	17.6
Penrith & Border	5.9
Stockton South	17.8
Westmoreland & Lonsdale	3.4

North West
Blackpool North	14.1
Blackpool South	13.8
Bolton North East	11.0
Bolton West	8.9
Bury North	7.0
Bury South	7.6
Chorley	6.8
Hyndburn	7.6
Lancashire West	11.1
Lancaster	7.6
Littleborough & Saddleworth	6.5
Morecambe & Lunesdale	12.7
Pendle	7.8
Ribble Valley	3.8
Rossendale & Darwen	6.5
South Ribble	6.2
Wyre	8.1

Yorkshire & Humberside
Batley & Spen	7.5
Beverley	5.4
Bridlington	8.7
Calder Valley	6.5
Colne Valley	6.6
Harrogate	4.7
Keighley	7.0
Leeds North East	9.0
Leeds North West	7.5
Pudsey	4.5
Richmond	5.2
Ryedale	4.4
Scarborough	9.3
Selby	7.0
Shipley	5.6
Skipton & Ripon	4.1
York	9.4

Greater London Labour-won seats

Barking	6.7	Islington South & Finsbury	12.4
Bethnel Green & Stepney	16.0	Lewisham Deptford	13.3
Bow & Poplar	14.2	Leyton	8.9
Brent East	10.7	Newham North East	9.1
Brent South	10.1	Newham North West	10.8
Dagenham	5.5	Newham South	11.0
Hackney North & Stoke Newington	14.1	Norwood	13.5
Hackney South & Shoreditch	15.5	Peckham	15.4
Hammersmith	11.3	Tooting	7.1
Holborn & St Pancras	13.2	Tottenham	13.7
Islington North	13.8	Vauxhall	13.9

Northern mean unemployment	8.1
Greater London mean unemployment	11.8
UK unemployment	8.7

(*Source*: Unemployment Unit, 1988.)

Table 4.32 Seats won by the political parties at the 1987 general election in the areas of lowest unemployment, Great Britain

	Unemployment rate in bottom decile (%)	Rank	Top-rank seat within bottom decile	Rank	Bottom-rank seat within bottom decile	Seats won C	L	A
Wales	7.2	354	Monmouth (C)	378	Brecon & Radnor (A)	2	0	2
Scotland	6.1	359	Orkney & Shetland (A)	474	Tweeddale, Ettrick & Lauderdale (A)	2	1	4
North	5.6	373	Carlisle (L)	560	Westmoreland & Lonsdale (C)	3	2	0
North West	4.9	432	Hazel Grove (C)	544	Ribble Valley (C)	7	0	0
West Midlands	4.7	455	Rugby & Kenilworth (C)	541	Stratford-upon-Avon (C)	6	0	0
Yorkshire & Humberside	4.6	470	Richmond (C)	533	Skipton & Ripon (C)	5	0	0
East Midlands	3.5	540	Rutland & Melton (C)	588	Harborough (C)	4	0	0
South West	3.0	532	Dorset West (C)	574	Dorset North (C)	5	0	0
East Anglia	2.4	601	Cambridgeshire South West (C)	625	Cambridgeshire South East (C)	4	0	0
South East	2.2	614	Windsor & Maidenhead (C)	633	Chesham & Amersham (C)	19	0	0
Regional bottom deciles						55	3	6
Great Britain (bottom decile)		570	Welwyn & Hatfield (C)	633	Chesham & Amersham (C)	62	0	0

Notes

C Conservative seats; L Labour seats; A Alliance seats.

(Source: Unemployment Unit.)

Table 4.33 Seats won by the political parties at the 1987 general election in the areas of highest unemployment, Great Britain

	Unemployment rate in top decile (%)	Rank	Top-rank seat within top decile	Rank	Bottom-rank seat within top decile	Seats won		
						C	L	A
North West	22.7	1	Liverpool Riverside (L)	18	Liverpool Mossley Hill (A)	0	6	1
North	20.7	2	Middlesbrough (L)	24	Newcastle upon Tyne North (L)	0	5	0
Scotland	20.6	6	Glasgow Central (L)	28	Glasgow Govan (L)	0	7*	0
West Midlands	18.4	5	Birmingham Small Heath (L)	66	Wolverhampton North East (C)	1	5	0
Yorkshire & Humberside	17.4	13	Sheffield Central (L)	51	Sheffield Brightside (L)	0	5	0
Wales	14.8	57	Cardiff Central (C)	86	Cardiff West (L)	2	2	0
East Midlands	14.6	21	Nottingham East (C)	153	Bassetlaw (L)	2	2	0
South East	13.2	47	Bethnal Green & Stepney (L)	221	Woolwich (A)	1	16	1
South West	12.6	55	Plymouth Drake (C)	210	Bristol West (C)	5	0	0
East Anglia	11.4	131	Great Yarmouth (C)	230	Norwich South (L)	1	1	0
Regional top deciles						12	49	2
Great Britain (top decile)		1	Liverpool (L)	63	Bradford West (L)	7	55	1

Notes

C Conservative seats; L Labour seats; A Alliance seats.

* At a by-election on 10 November 1988, the Scottish National Party won Glasgow Govan – formerly a Labour-held seat.

(Source: Unemployment Unit.)

of seats in rural areas (for example, in the Cheshire Plain, north Lancashire, the East Riding and the Vale of York, and in the Welsh Marches), and also represented the affluent suburbs of large industrial cities (such as North East Leeds and Sheffield Hallam), resort and retirement centres (for example, Blackpool and Harrogate) and even some industrial towns (such as Bolton and Darlington). But while the Conservatives were able to win a solid wedge of 37 seats across the whole of the north of England from west to east in 1987, Labour retained a block of seats in the capital, 23 chiefly in inner London.

As Table 4.31 indicates, it is notable that the level of unemployment in the Conservative wedge of northern constituencies was significantly lower than in the Labour seats of London – a reflection perhaps of satisfaction/optimism in many parts of the north and dissatisfaction/pessimism in much of the capital. The level of unemployment, as opposed to location *per se*, might thus be seen as the major determinant of voting behaviour. With regard to Britain as a whole, evidence to support this claim is obtained from an examination of the electoral performance of the two main parties in the areas of lowest and highest unemployment within each region in 1987.

Table 4.32 thus shows that bottom-decile constituencies in all regions (those with the least unemployment) generally had more in common with each other (they largely returned Conservative members) than with the rest of the constituencies within their region. Conversely, as Table 4.33 shows, top-decile constituencies (those with most unemployment) broadly voted the same way (mainly returning Labour members) irrespective of voting behaviour elsewhere in their region.

From the brief analysis above, it is clear that there is no simple north-south divide in political support: the Conservatives hold a sizeable part of the north of England while Labour retains a major foothold in London. The level of unemployment (as an indicator of local optimism/pessimism), rather than an area's position on a map, clearly has a substantial influence on voting behaviour.

Notes

1. Fothergill, Kitson and Monk (1987) assumed that if manufacturing output were to increase by 3.5 per cent per annum (in the short-to-medium term), if this necessitated a proportionate increase in floor-space each year, if demolitions and factories no longer suitable for use were taken into account, and if half of an estimated 190 million m^2 of required floor-space were needed in new units (as opposed to refurbished property), then the total derived demand for industrial land would amount to 31,600 hectares – an amount of land probably less than the total availability of industrial sites.

2. In research commissioned by the Department of Trade and Industry, Tyler, Moore and Rhodes (1988) showed that input unit-costs tended to be higher in large conurbations than in their surrounding hinterlands. This was particularly the case in London, Birmingham, Liverpool, Newcastle upon Tyne and Sheffield (although in Leeds and Manchester the dominant low-wage textile industries dampened local cost structures). The conurbations were also shown to be adverse locations in terms of productivity. This counters the view that the high costs of production generally incurred in cities can be more than offset by above-average levels of productivity. Tyler, Moore and Rhodes, therefore, concluded that, with regard to both costs and productivity, the conurbations are now at a competitive disadvantage compared to their hinterlands.

5

POLICY FOR THE 1990s

It is very evident from an examination of interregional variations in employment growth, unemployment, incomes and expenditure, population growth, housing, health, the 'quality of life' and voting behaviour that there is a very marked north-south divide, and that under successive Thatcher administrations the divide has worsened. In the view of Massey (1988), central government in the 1980s manifestly championed 'the cause of certain already-privileged groups' in society, and particularly those with comparatively high levels of incomes and accumulated capital assets overwhelmingly concentrated in the South East. It is equally clear that within London, adjacent to areas of considerable wealth creation, there is an appallingly high incidence of deprivation (including high unemployment); and that intra-regionally, as well as interregionally, there are stark disparities in prosperity – particularly between the inner cities on the one hand, and the outer suburban and country areas on the other.

The north-south divide: policy considerations and options

In the late 1980s, it was apparent that north-south divisions were likely to widen still further. On the assumption that the current political and economic scenario continued into the early 1990s, Gudgin and Schofield (1987) forecast that, although growth in national output would be sufficient to allow expansion in all parts of the UK, this would neither be the case for employment nor population. Large parts of the north (due to the continuing decline of 'traditional' manufacturing) would lose rather than gain jobs over the period 1987–2000, while employment in the private-services sector would continue to increase in the south; and in the same period (because of interregional migration), the population of the north is likely to decline while the population of the south is set to increase. In short 'the future contrast is likely to be between a booming southern half of the country and a slowly declining northern half' (Gudgin and

Schofield, 1987). In the 1990s, moreover, whereas it would require a rapid growth in the world economy and extremely effective regional policies to reduce the level of unemployment to below 10 per cent in all northern regions, a further period of recession (possibly associated with high rates of interest) would have an immediate and adverse effect on the north, since companies (as in the period 1979–82) would close down their branch plants in northern locations (Gudgin and Schofield, 1987; Fothergill, 1988).

It is sometimes posited that there is little merit in attempting to regenerate the economy of the north since market forces would inexorably ensure that resources would be more profitably employed in the south. Instead (or so it could be argued), governments should plan the economic decline of the north as humanely as possible – for example, by spending discriminatorily on environmental improvement and on social welfare but eschewing any idea of industrial assistance. This view, however, must be rejected in its entirety. It would not only be against the economic interests of the north, but would also be detrimental to the south and the national economy. In general terms, it is very probable 'that if economic growth in the United Kingdom continues to concentrate in the southern part of the nation this will ultimately act as a constraint on the overall growth of the national economy, and thus the volume of resources available to all, northerner or southerner' (Tyler, 1987). It was evident that a consumer boom in the mid- and late 1980s (with a resulting high rate of inflation and excessive import expenditure) was chiefly associated with overheating in the south, while a comparatively low level of manufactured exports was, in large measure, a result of continuing deindustrialization in the north.

Past attempts to disinflate through high rates of interest/reduced levels of public expenditure have had adverse effects both on the rate of economic growth in the UK as a whole and on the vulnerable economy of the north in particular, and there is no reason to suppose that broadly similar macroeconomic policies would not have the same effects in the late 1980s to early 1990s.

Clearly, as an alternative, economic regeneration in the north would not only reduce inflationary pressures in the south and in the economy as a whole, but would also help to correct adverse balances of (non-oil) trade. Since both the Department of Trade and Industry (1985) and House of Lords (1985) had claimed that the weakness of the manufacturing trade balance had serious consequences for the vitality of the national economy, it would have been very appropriate (when faced with a rising inflation and record balance-of-payments deficits in 1988 and 1989) to have embarked upon a policy to reindustrialize the manufacturing base of the north (Martin, 1988).

The reindustrialization of the north

While the north requires both a regeneration of manufacturing and a better performance from the service sector, it is severely disadvantaged by a poor rate of new-company formation, a deficiency of 'latent indigenous potential' for business expansion and a greater reliance on relatively low-wage public-sector employment (rather than higher-earning private-sector activity), perpetuating 'the divide between the low paid north and the high paid south' (TCPA, 1987, p. 16). This divide, argues Massey (1988), maintains a 'huge rigidity in the labour market', with migration from areas of labour surplus in the north to areas of shortage in the south being impeded by substantial house-price variations – the continuation of labour shortages in the south inflating wages (and house prices) still further. Massey points out that the divide is also manifested by underused infrastructure in the north (for example, motorways) and congested infrastructure in the south – necessitating higher levels of public expenditure particularly in the South East. Clearly, because of imbalance, 'we are writing off already paid for investment in the north and paying for more in the south – and all of us pay' (*ibid.*). It is evident, therefore, that the full private and social costs of economic decline in the north and congestion in the south are not taken into account in location decision-making in both the private and public sectors.

Robson (1987) suggests that central to the creation of a regionally balanced labour-market is the need to attract corporate headquarters to the north; to establish a strong regional dimension to business enterprise (possibly through a process of corporate devolution or de-merger); to develop a number of 'northern magnets' in which information-rich and intelligence-rich high-technology environments could be created; and to undertake transportation and environmental improvements. If the private sector is to perform leading roles in the regeneration of the north, it is essential that government, as a prerequisite, both tackles private capital's reluctance to invest outside the south, and introduces positive incentives to reverse the slide of capital from the north to the south (see Halsall, 1988).

Inappropriate and impracticable policies

It is sometimes argued that, since national economy recovery has reduced regional disparities in the labour-market at numerous times in the past, the most appropriate way for government to narrow current disparities would be to reflate the economy by raising the level of aggregate demand (see Armstrong and Riley, 1987). However, in the 1980s, general reflation would neither have had a major effect on unemployment in the north, nor would have reduced interregional disparities in the labour market. As Armstrong and Taylor (1987a) pointed out, the slump of 1979–82 had resulted in 'the premature scrapping of productive capacity in manufacturing', and therefore the subsequent recovery in output in the mid-1980s consequently led to manufacturing industry operating at full capacity but at a markedly lower level of both production and employment than in the 1970s. Indeed, it could even be argued that reflation in the form of cuts in income tax (such as those introduced by the Budget of 1988) had harmful effects on the economy of the north. Higher consumer spending (rather than benefiting manufacturing industry in the north), sucked in imports, boosted the rate of inflation and created further shortages of skilled labour in the south – widening interregional disparities in earnings. Ultimately, higher interest rates in 1989 (aimed at curbing inflation) almost certainly damaged the fragile northern economy rather than significantly deflating the overheated south (except in the case of house prices).

An alternative approach to the problem of a disparite labour-market, and one favoured by free-marketeers, would involve the replacement of national wage-bargaining by a system of local or regional wage determination. In the north, where the supply of labour exceeds demand, people would thus 'price themselves into a job' and consequently attract industrial development from the south or from overseas. But in the view of the TCPA (1987) such a proposal would be both ill advised and impracticable and must be rejected. In order to compete successfully with major overseas manufacturers both domestically and internationally, the whole of the UK (argued the TCPA) would need to price itself into markets and this might necessitate labour in the south, as well as in the north, taking substantial wage cuts. Whether confined to the north or adopted nationally, a policy of wage cuts would be strongly resisted by the trade unions and stand little chance of being implemented on any scale. Wage cuts, moreover, would reduce the level of aggregate demand, and through the reverse multiplier reduce the level of employment in those activities dependent on local spending power (TCPA, 1987).

It is often argued that interregional disparities in the labour-market could be largely eliminated if most of the unemployed in the depressed regions migrated to jobs in the south. Smallwood (1988), for example, suggested that this could reduce unemployment in Great Britain by as much as half a million and that, because of an increased labour supply, wage inflation in the south would be largely countered. Because council tenants find difficulty in migrating regionally since it is necessary to secure an exchange of housing, and while owner-occupiers in much of the north normally find house prices in the south unaffordable (except at a substantially higher level of income and mortgage), it might be assumed that geographical immobility would be eased if an adequate supply of private rented housing could be ensured. Partly to this end, the Housing Act 1988 (by permitting landlords to charge market rents on new lettings) aimed to reduce shortages in the private rented sector and to facilitate a greater provision of low-income housing by housing associations and co-operatives. It is highly questionable, however, whether these provisions 'measure up to the scale of the labour mobility problem or constitute the most effective immediate response to it' (Smallwood, 1988). But access to owner-occupied housing in the south is possible where companies assist with mortgages or purchase a share in housing, and where relocation consultants or the Confederation of British Industry's Employee Relocation Council advise on moves (Halsall, 1987).

However, as a solution to labour-market disparity, large-scale migration to the south would be disadvantageous to both the north and recipient areas. Armstrong and Taylor (1987a) strongly argue that it 'would starve the north of young, skilled, well qualified [and] adventurous labour' – a crucial resource needed in the north for its regeneration, while the south would suffer increasing pressures on its housing market, infrastructure and public services.

Discriminatory public expenditure

If, for the reasons previously examined, it is necessary to minimize interregional disparities in prosperity it is essential to employ regionally discriminately fiscal measures. First and foremost, cuts in regional spending (undertaken in the late 1970s and throughout the 1980s) must be reversed. Since regional policy had, in general, enjoyed considerable success in the post-war decades (see Moore, Rhodes and Tyler, 1986), there is clearly a strong case for a return to mandatory regional development grants (or their equivalent), the reintroduction of Special Development Areas where unemployment is substantially above the level of the UK as a whole, and the designation of more parts of the country as Development Areas. Apart from inducing a greater volume of investment in the Assisted Areas from domestic funds, a strong regional policy might be necessary to deflect inward investment to the north (notwithstanding the unpredictable long-term impact of investment from overseas on the UK economy). Whereas the winding down of aid in the Assisted Areas did not deter Nissan from locating in Sunderland in 1986 or Fujitsu and Bosch from opting in 1989 to locate respectively in Newton Aycliffe (Co. Dur.) and Miskin (Mid Glam.), Toyota might have decided (in 1989) to have located its £700-million, 3,000-employee car plant in an Assisted Area rather than near Derby in an area of relatively low unemployment – had regional aid been more attractive.

By the late 1980s it was evident that the UK's macroeconomy had significantly recovered from the recession of 1979–82, but there were still unacceptable peaks of unemployment throughout extensive areas of the north. Given existing regional policy, it was highly probable that unemployment would rise again if inflation resulted in the introduction of insensitive deflationary measures, such as high interest rates. Apart from 'traditional' regional aid, many other forms of public expenditure should be spatially discriminatory. Funding incentives to high-technology industry, for example, must be targeted at the Assisted Areas since purely national incentives to encourage innovation would merely result in further overheating in the south (see Oakley, Rothwell and Cooper, 1989). Likewise, while a strong case can be made for redirecting arms expenditure to more socially useful production, since defence-equipment spending continues to make a major claim on Exchequer resources, it should cease to be very largely concentrated in the south and be distributed more extensively throughout the regions (see Boddy, 1987). Similarly, the armed forces (disproportionately based in the south) should decentralize (*The Economist*, 1988).

Environmental expenditure, too, could be biased towards the north. Undoubtedly, the restoration and improvement of the landscape (including urban renewal) and the provision of leisure and recreation facilities could enhance the appeal of many parts of the north to investors – not least to major firms planning the development of branch plants or even the relocation of their headquarters.

Improved communications (particularly air and rail services) are, of course, vital for the regeneration of the northern economy. Rather than extending airports in the South East to handle an ever-increasing volume of traffic, the number of international flights in and out of northern airports should be increased. It is notable that while Manchester Ringway airport, for example, operated at well below capacity in 1988 – accommodating a throughput of only 10 million passengers – the London airports of Heathrow and Gatwick in the same year were increasingly congested, carrying respectively 37.5 million and 20.8 million passengers – many of whom had either travelled from or been *en route* to the north. Since accessibility to a major airport is an important locational consideration for many businesses (see Hall *et al.*, 1987), it is important therefore that 'the government should use its licensing and other powers to ensure that regional airports enjoy an improved level of services with more direct international connections' (TCPA, 1987, p. 18). As the TCPA pointed out, since Manchester Ringway – of all regional airports – has the greatest scope for expansion, it should be extensively developed as an alternative to Heathrow/Gatwick/Stansted and could provide a frequency and 'range of international scheduled services comparable to other major European airports' (*ibid.*). Clearly, it would be as much in the interests of the south as of the north that both air and motorway congestion in the vicinity of London's airports was reduced by the permanent diversion of a significant proportion of international air-traffic to Manchester.

With regard to rail services, it is very probable that a failure to provide a new rail corridor north of London would not only result in a high proportion of northern exporters having to continue to rely on comparatively costly road transport to the Eurotunnel shuttle at Ashford (Kent), but also having to suffer increasing delays as motorway links (particularly the M25) became more and more congested (Gossop, 1989). Exporters in the South East, in contrast, would have greater accessibility to the Channel link, and thus have a considerable advantage over their northern rivals. The failure to develop a direct rail service between northern cities and the European Continent would also impose constraints on passenger demand: adversely affecting business contact between northern-based firms and their international markets. There is, therefore, not only the urgent need to invest substantially in adapted and/or dedicated track from the Channel tunnel to the northern regions (with links around London), but also to develop trans-shipment depots and private sidings at key locations to ensure that the north can compete effectively with the South East on the Continent of Europe (*ibid.*). However, by the end of the 1980s, few of these essential developments were underway or even programmed.

It is of concern that, instead of investing heavily in an extended Channel rail-link to the north, the government will embark upon a major roads programme in the 1990s (see Department of Transport, 1989). Apart from transforming the M1 and the M6 into eight-lane super highways, a network of extended motorways will link the North East, Yorkshire & Humberside, the North West and the West Midlands to the M2 and the Channel tunnel. While the long-term overall comparative costs and benefits of rail and road development and use may be uncertain (on environmental grounds alone, rail transport is undoubtedly less costly than motorway use), a principal and well-acknowledged criticism of motorway development in the late twentieth century is that additional road space is soon used to full capacity with a resulting escalation of both private and social costs. If this reduces the ability of British manufacturers to compete within the single European market, the north-south divide would be widened rather than narrowed by the extension of the motorway network. It would therefore seem economically unwise for highway investment to take precedence over rail investment in the 1990s.

The narrowing of interregional disparities in education, housing and health is also necessary. This would not only improve the quality of life in many parts of the north but would also almost certainly help to eliminate imbalances in the labour-market. Poor education is manifestly associated with shortages of skilled labour and high unemployment; social *malaise* may be attributable to inadequate housing (both quantitatively and qualitatively) and a comparatively low life expectancy may be significantly linked to bad diets, smoking and alcohol abuse (see Haskins, 1987). Discriminatory public expenditure in areas of the north on education, house-building and rehabilitation, and preventative medicine would do much to enhance the marketability of labour in growth sectors of the economy.

The impact of expenditure on public administration has historically been greatest in London; therefore, the decentralization of government can clearly be regarded as a form of regional aid (and as a means of reducing congestion in the capital). Whereas in the 1970s it was public policy to decentralize government offices away from London to stimulate the creation of new, ancillary service industry in the recipient regions and to encourage private firms in the South East similarly to relocate, in the late 1980s there were, in addition, very significant cost advantages to the public sector in moving out of the capital. As *The Economist* (1988) pointed out, many government departments in London find 'it difficult to attract and retain staff at competitive pay rates'[1]; every department (whether or not its buildings are government owned) is charged market rents and these, of course, are considerably higher in London than in the north; and bigger local-pay variations (introduced in the late 1980s) will increase the savings from relocating away from London. In 1989 (chiefly due to these considerations), it was announced that 2,000 jobs from the headquarters of the Departments of Social Security and Health would be moved to Telford, Cardiff and Milton Keynes, and a further 1,000 jobs in benefit offices would be dispersed from the capital to locations within the Assisted Areas (for example, in Glasgow and Wigan). Even local authorities in London, in the view of *The Economist*, could decentralize much of their administrative and clerical work. Since an increased volume of service activity in parts of the north could contribute significantly to regional development, it is important that initiatives to decentralize government-office employment are, at the minimum, maintained.

Overall, the full regional impact of government expenditure must be estimated in broad cost-benefit terms and published annually. Such a regional audit (commonly used in other advanced capitalist countries such as the USA and Canada) would apply not only to regional aid and industrial innovation, but also to defence, environment, transport, education, housing and health. Public expenditure could subsequently be linked to a quantitative needs-index to discriminate both positively and precisely in favour of the northern regions (see Armstrong and Taylor, 1987a; Boddy, 1987).

Discriminatory taxation

If a policy of positive discrimination were to be extended to taxation, it would be necessary, as a major priority, to ensure that new forms of local taxation were regionally progressive. Whereas the introduction of a single uniform business rate (in England and Wales) in 1990, under the Local Government Finance Act 1988, will probably raise rate payments on offices and shops in the south but lower the rate burden on industrial properties in the north – with businesses in general losing in the south but gaining in the north (see Balchin, Bull and Kieve, 1988) – the impact of the accompanying community charge will be very different.[2] Fleming and Nellis (1989) have shown that, whereas there will be a large number of net gainers in the South East and West Midlands, the number of net losers in the North and in Yorkshire & Humberside will be quite considerable (Table 5.1). However, they suggest that since domestic rates (as a percentage of the net housing expenditure of owner-occupiers) had been generally higher in the north than in the south (Table 5.2), the introduction of the community charge will inflate the demand for housing at a disproportionately greater rate in the north (although not in the North region and in Yorkshire & Humberside where the charge will create more losers than gainers). While percentage increases in house prices might be much greater in the north, absolute price increases are likely to continue to be more substantial in the south – widening the north-south divide. Table 5.2 shows that the difference in house prices in 1988 between, for example, the North West and Greater London was £50,163, but if – in response to the abolition of rates – real house prices in the North West increased by, say, 20 per cent but rose by only 12 per cent in Greater London, the house-price difference between the two regions would increase to £53,407.

Table 5.1   Estimated regional impact of the community charge in England

	Thousands of households			Percentage of households
	Losers	Gainers	Net gainers or losers	
South East	1,080	2,970	+ 1,890	73.3
West Midlands	610	1,540	+ 930	71.6
East Anglia	295	543	+ 250	64.8
South West	710	1,010	+ 300	58.7
North West	1,130	1,450	+ 320	12.4
East Midlands	780	900	+ 120	7.1
Greater London	1,185	1,305	+ 120	4.8
Yorkshire & Humberside	1,325	875	− 450	20.5
North	850	550	− 300	21.4
England	7,970	11,145	+ 3,175	58.3

Note
The table shows the number of gainers and losers after the community charge has been phased in over three years.

(*Source*: Fleming and Nellis, 1989.)

Table 5.2 Domestic rates in relation to owner-occupier housing expenditure and house prices in Great Britain

	Net rates as a percentage of net housing expenditure for owner-occupiers, 1985–6	Average house prices, 1988*
		(£)
Scotland	27.5	31,927
North	22.6	29,671
North West	20.8	34,690
Yorkshire & Humberside	19.9	37,573
West Midlands	19.1	47,333
East Midlands	18.1	46,075
Greater London	17.8	84,853
South West	16.2	68,046
South East (excl. Grt London)	15.9	79,953
East Anglia	15.6	62,622
Wales	13.7	36,311

Note
* Fourth quarter, 1988

(*Sources*: Fleming and Nellis, 1989; Building Societies Association, *Housing Finance*.)

Since the largest number of net gainers from the introduction of the community charge, and the greatest absolute increase in house prices, are likely to be in the South East, a policy of positive discrimination would necessitate a net northward flow of needs subsidy to facilitate the imposition of a relatively low community charge in areas of net loss. Alternatively, the community charge could be replaced by another form of local taxation, for example, a local income tax or a site-value rate, accompanied by a system of distributing revenue from high-yielding to low-yielding areas (see Balchin, Bull and Kieve, 1988).

Under a discriminatory tax policy, tax relief must be adjusted or phased out to eliminate disproportionate benefits to the south. In the late 1980s, whereas service industries (located mainly in the south) notably benefited from a lower rate of corporation tax, manufacturing industry (disproportionately located in the north) was particularly disadvantaged by lower capital allowances. Likewise, mortgage interest relief (amounting to over £5,000 million by 1989) overwhelmingly favoured owner-occupiers in south. If corporation-tax allowances are to remain, there is clearly a need to divert their flow to the north rather than concentrate assistance in the south; and whereas there are many calls for the complete abolition of mortgage interest relief, not least from the National Federation of Housing Associations (1985), the retention of relief (or preferably the introduction of tenure-neutral housing allowances), inversely geared to income and positively related to housing costs, might do much to even out the regional impact of housing subsidies.

There is a view, however, that more radical tax reform is necessary. Mr Michael Heseltine, MP (1987), argued that capital taxation in the corporate sector should cease since this would neutralize the ability of London-based public-quoted companies to offer tax-deferred shares to acquire businesses in the regions; and that pension funds should no longer enjoy the benefits of tax-free income that enables them to attract a substantial share of the nation's savings only to invest it disproportionately in southern-based companies and financial institutions. A more discriminatory reform, however, would be to retain capital taxation fully but amend tax privileges to favour investment in the north rather than in the south - possibly through the creation of regional capital-venture funds managed by the government and the institutions in partnership.

The full regional impact of taxation must (as far as possible) be estimated and, together with a consideration of the regional impact of public expenditure (see above), help to guide the formulation of discriminatory fiscal policy.

It is possible, however, that the volume and allocation of regional aid in the UK will be increasingly determined within the context of the European Community (EC). At the Brussels' summit in February 1988, it was agreed that spending on regional development would be doubled in the years preceding the establishment of a single European market in 1992. The UK was expected to contribute about 15 per cent of the total £36 billion budget, while the increased volume of assistance would be distributed through the existing Structure Funds, in other words, the European Regional Development Fund (ERDF), the European Social Fund and a section of the European Agricultural Guidance and Guarantee Fund. National governments (on being granted aid) will be required to contribute up to 50 per cent of the total assistance available or 30 per cent in special (privileged) cases. Whereas, in principle, similar percentages were sought in the past to ensure that ERDF assistance was additional to national funding, the UK government in the 1980s frequently clawed back its planned expenditure on regional aid and relied wholly on EC funds to finance projects that it otherwise would have supported – with the flow of resources to the north being consequently about half of what was intended under EC law. Under the community's Single European Act 1986, however, powers have been conferred on the European Commission to prevent national governments from, in effect, creaming off EC resources destined for the deprived regions.

In December 1988, the Commission of the European Communities (CEC) declared that (as from 1 January 1989) the amount of annual aid available under Structure Fund provisions would increase from £5.5 billion to £11 billion; and in March 1989 it was announced that 33 areas in the UK (containing a total population of 20 million) would become eligible for a share of the £21 billion of ERDF aid available under *Objective 2* criteria, including many parts of the country that did not currently qualify for national regional assistance. In terms of population, the UK received the lion's share of eligible areas, around twice that of any other member state. Within these areas, the CEC will fund local schemes to generate employment, improve infrastructure and facilitate training. If regional disparities are to be reduced within the UK, it is clear that (short of radical political/economic change), an increasing dependence upon EC funding will be necessary.

Policies for the inner cities

In a mixed economy it would seem inappropriate to rely exclusively either on private enterprise or the public sector to provide solutions to the many and complex socio-economic problems of the inner cities. Indeed, during the 1970s and 1980s, it was recognized by successive governments (and not least by the Thatcher administrations) that a partnership between both sectors of the economy was fundamental to urban regeneration. By the late 1980s, the Department of the Environment was spending over £500 million on urban aid (see Table 3.6), a substantial proportion of which levered private investment into inner-city projects – often by a ratio of at least 1:4. Rather than promulgating a free-market economy, Thatcher governments increasingly intervened in, and distorted, the market through establishing such agencies as urban development corporations, city-action teams and task forces, and by means of enterprise zones and urban renewal grants of various sorts (see Chapter 3).

As experience in the USA has shown, the real hope for the inner cities lies in a joint public–private-sector approach, but in the UK during the 1980s the impact of partnership between the two sectors has been severely weakened by the ideology of Thatcherism. First, an attachment to monetarism throughout much of the decade led to a substantial reduction both in the rate-support grant and in housing subsidies to local authorities, often dwarfing urban aid; and second, central government increasingly isolated local government from investment decision-making at a local level – thereby eliminating local accountability and the likelihood of development being targeted to meet local needs (particularly within the field of employment and housing). Thus, an increasingly centralist government was not only (in net terms) withdrawing funds from the inner cities, but – in effect – was also attempting to bring about the collapse of the local state. As the TCPA (1987) observed, Whitehall has become more and more powerful, while (in contrast) in West Germany, 'the federal economy has flourished on the basis of strong, autonomous provincial cities' (p. 6), and in the USA there has always been a system of great regional centres – Boston, Atlanta, Chicago and San Francisco. Critics of Thatcherite policy might argue that instead of squeezing out local authorities from the urban regeneration process, central government should exploit fully the expertise and the current initiatives of local government to develop the assets of the inner cities.

As an alternative to 'top-down' solutions, local authorities in the 1980s developed their own regeneration strategies. A number of local enterprise boards (LEBs) were set up to provide 'seed money' for local ventures (in the hope of attracting consequential institutional investment), equity and loan guarantees (to enable firms to borrow longer term and for more risky venture than would otherwise be possible), and delayed repayment loans. LEBs, for the most part, were resourced from the product of a 2p rate-levy under Section 137 of the Local Government Act 1972, but potentially had at their disposal their local authorities' pension funds. However, while five major LEBs were set up in the 1980s (the Greater London Enterprise Board, the West Midlands Enterprise Board, the West Yorkshire Enterprise Board, the Merseyside Enterprise Board and Lancashire Enterprises Ltd) and, although each were able to create jobs at a significantly lower level of investment than that undertaken within the enterprise zones or through the medium of urban development corporations, many district authorities preferred to promote local industry directly – usually through their economic-development departments. Sheffield, for example, used the product of a 2p rate to finance local co-operatives, and applied a policy of positive discrimination in favour of local firms in placing procurement contracts (Balchin and Bull, 1987); Rotherham concentrated on townscaping and demolishing eyesores as a means of attracting enterprise; while Leeds hoped to attract investment by restructuring its inner-city areas through the development of a science park, by support for its technologically oriented university, by the planting of more trees per capita than any city authority in Europe and by the creation of the largest pedestrianized shopping centre in Britain (Breheny, Hall and Hart, 1987). Compared with the total amount of unemployment in the inner cities, however, local-authority initiatives have not generated many jobs but, as Blazyca (1983) pointed out (in referring to LEBs), they have shown that there is an alternative to monetarist inner-city policy. With more funding and more co-ordination (possibly within a regional-planning framework), LEBs and their like might be capable of producing more effective urban strategies than those employed by successive governments in the past.

While urban-regeneration policies might be designed to create wealth, the processes of urban renewal often have a regressive effect on the distribution of wealth and inflict external costs on the urban environment. Improvement grants awarded under the provisions of the Housing Acts 1969, 1974 and 1980 have attracted an inflow of middle- and high-income households who have eagerly purchased rehabilitated houses or flats (or properties ripe for improvement) – either with the intention of reducing travel costs and travel time to work or of realizing speculative gains (or possibly both). Gentrification has often been associated with the displacement of low-income tenants (usually by landlords or developers during the initial stages of refurbishment), with homelessness and associated poverty resulting (see Merrett, 1976; Balchin, 1981; Massey, 1988). Similarly, both finance and industrial capital have been attracted into the London Docklands, where massive residential and commercial developments have done very little to ease the housing and employment problems of the London boroughs of Tower Hamlets and Newham. As Massey (1988) pointed out (both in the case of gentrification and dockland redevelopment), subsidies have often benefited national or multinational capital rather than the local entrepreneur or local capital. Likewise, in the north, although a number of towns compare very favourable with their counterparts in the south in terms of their environment and their growth potential (see Breheny, Hall and Hart, 1987), this should not detract attention away from the adverse effects of insensitive development in historic cities such as York and Durham, or from the very urgent need to regenerate the economies of the major industrial cities of the north. Within urban Britain in the 1990s, what is at stake, therefore, is not how the effects of economic decline can best be cushioned but how an efficient and equitable renewal policy can be formulated and applied. Clearly, 'in the end it is the government's model of growth, and its class politics, which are at issue' (Massey, 1988). Of overwhelming importance is the conundrum of whether or not the inner cities have priority over the regions in respect of claims on public expenditure.

Increasingly throughout the 1980s, the government believed that the inner cities did have first claim, and resources were allocated accordingly.

While government should look beyond the inner cities and formulate 'distinctive regional policies which recognize the extent of the imbalance between different parts of the country' (Brittan, 1987a), it must be recognized that there are very significant economic disparities intra-regionally as well as interregionally. Without an intra-regional dimension, future regional policy would not only lack the sensitivity to tackle the very real socio-economic problems that exist within all regions of Britain, but would also be incapable of facilitating the aims of macroeconomic policy. It is for this reason that we turn now to the question of regional government.

Devolution to the regions

While the north-south divide, in aggregate, has been getting wider, political and economic power has become increasingly concentrated in London. The capital, overwhelmingly, has become not only the centre of government but also the hub of commercial and financial activity in the UK, and the focus of air, rail and road transport. In contrast, local-authority responsibilities throughout Great Britain have been seriously eroded in recent years, while firms within the north have been taken over by national or multinational concerns with headquarters more accessible to the centre of power (see Luttrell, 1987). Marked regional inequalities combined with the emergence of what is virtually a unitary State are problems that no other western European country has to face. Clearly, regional imbalance in both prosperity and power can only be significantly reduced if Britain (like its neighbours within the European Community) becomes a country with several regional centres and not just one capital.

The centralization of political power, however, is not synonymous with efficient government - regardless of whether or not the narrowing of interregional disparities is a central plank in public policy. Currently (with the obvious exception of decision-making within the Scottish, Welsh and Northern Ireland Offices), policy determination in all central-government departments is concentrated in London but (in terms of their regional dimension), policies are often - at best - poorly co-ordinated or - at worst - seriously contradictory. There is clearly a need to 'break with traditionally compartmentalized views of the role of the state in regional and urban development (Damesick and Wood, 1987, p. 264) - particularly within the Departments of Trade and Industry, Transport and the Environment. Damesick and Wood suggest that departmental policy responses to the problems of regional and urban development (apart from being better co-ordinated) need to be 'set within a framework of regional strategic thinking and planning' (*ibid.* p. 266), and that more effective channels of communications should be established to enable the regions 'to press their claims more forcefully upon central government' (*ibid.*).

Critics of devolution often argue that the setting up of regional authorities would create a further (and unnecessary) tier of government. However, if there is a strong case for establishing some form of regional administration (and evidence suggests that there is), it is necessary to consider whether or not the current system of local government should remain unchanged. Both the Conservative-controlled Association of District Councils (ADC) and the Labour Party favour the abolition of county councils and a consolidation of local government below the regional level. The ADC suggested that since the government's programme for planning, housing, social services and the community charge will impose a very great burden on the county councils in the 1990s (in part due to their remoteness from local communities), the case for transferring these services (and much of their funding) to a single tier of district councils will prove irresistible (see Hughes, 1988a); and the Labour Party (1989, p. 58) reaffirmed its belief in the need to dissolve the county councils and create 'most-purpose local authorities based wherever appropriate on existing districts'. While the government – in the late 1980s – was opposed to abolishing the county councils, it nevertheless was committed to reducing their powers. The white paper, *The Future of Development Plans* (Department of the Environment, 1989), proposed that structure plans (which, since 1968, had been drawn up by the counties) should be abolished, leaving the district authorities sole responsibility for planning at the local-government level. Without their powers to plan, the county councils would find it difficult to justify their continued existence.

Prior to an examination of proposals to devolve powers from central government to the regions, a brief consideration of economic development in Wales will show first how a form of regional government has – to a degree – regenerated a depressed region; second, an examination of the South East of England will explain why regional planning and devolved government might be necessary even in a seemingly prosperous region; and, third, a look at some of the aspects of regional economic development in Continental Europe will suggest ways in which institutional change in the UK might facilitate the creation of a more balanced economy.

Lessons from Wales

Wales is one of the least prosperous regions of the UK. Across a wide range of economic and social indicators, the principality compares very unfavourably with the more affluent parts of the country and particularly the South East (Table 5.3). Despite a long period of deindustrialization and the prospects of further decline in the foreseeable future, a clear majority of the Welsh electorate at the 1979 referendum rejected proposals to set up devolved government. In the principality there was clearly a strongly held belief that the economic and political future of Wales was bound inextricably to the UK as a whole, but this conviction must have seemed increasingly difficult to maintain throughout the early 1980s when interregional disparities were widening.

Following the return of a third Thatcher administration in 1987, Mr Peter Walker, MP, was appointed Secretary of State for Wales. He subsequently attempted to demonstrate that 'a non-Thatcherite, one-nation brand of Conservatism can work' (Hetherington, 1987). When major public-expenditure cuts were being considered in the autumn of 1988, he stressed the need to at least

Table 5.3   Wales and the South East compared

	Wales	South East
Total employment change, %, 1979–86	− 13.0	+ 2.0
Manufacturing employment change, %, 1979–87	− 34.5	− 27.3
Service employment change, %, 1979–87	− 1.6	+ 11.5
Business registrations change, %, 1979–87	+ 11.8	+ 23.9
Unemployment, %		
1979	6.5	3.0
1986	14.9	8.6
1988	10.9	5.5
Vacancies as a % of unemployed, 1988	7.5	23.9*
Gross domestic product per capita (GDP) at factor cost, 1987	£4,989	£7,183
% change in GDP at factor cost, 1983–7	30.6	42.9
Total personal incomes per head, 1987	£5,144	£7,212
Personal disposable income per head, 1987	£4,138	£5,560
Estimated population change, %, 1981–7	+ 0.8	+ 1.8
Estimated net migration, %, 1981–7	+ 0.5	+ 0.6
Labour-force with a degree, %, 1986	8.1	12.6
Managerial and professional groups as a % total employees, 1986	11.1	17.1
House prices, average, 1989 (first quarter)	£47,110.00	£99,629.00 OMA £91,783.00 GL £85,035.00 OSE
Market value of the housing stock, % change, 1981–7	+ 74.00	+ 177.00
Death rate per thousand:		
Men aged 20–64	5.68	4.88
Single women aged 20–59	1.43	1.29
Married women aged 20–59	2.34	1.97

Notes: OMA Outer Metropolitan Area;   GL Greater London;   OSE Outer South East.
* Excludes Greater London.

(*Sources*: Department of Employment; Central Statistical Office; Office of Population Censuses and Surveys; Department of the Environment; Nationwide Anglia Building Society.)

maintain the current level of regional aid – despite a worsening balance-of-trade deficit and a higher rate of inflation. He argued that it was precisely at times like these that industry and commerce must be attracted to Wales and to the other depressed areas of Britain, and that – in addition – regional aid was necessary to attract foreign investment into the Assisted Areas to the long-term benefit of the trade balance. Without this inducement, investment from Japan and the USA would overwhelmingly take place in Continental Europe prior to the opening of the single market in 1992. Mr Walker also recognized that a high level of public expenditure was required in Wales to clear up the legacy of environmental pollution and to fund the retraining of redundant labour, such as ex-miners (see Lewis, 1988). For all these reasons (and undoubtedly in an attempt to increase Conservative support in the principality), he was able to secure a 36-per-cent increase in the regional-aid budget for Wales from £156 million in 1987-8 to £212 million in 1989-90, including an increase in the funding of the Welsh Development Agency (WDA) from £57 million to £73 million (Table 5.4).

Mr Walker, however, had inherited a regional economy that was already recovering from the recession of the early 1980s when he replaced Mr Nicholas Edwards, MP, as Welsh Secretary. In 1987, the gross domestic product (GDP) per capita of the principality was 15 per cent greater in money terms than in 1985, whereas it had grown by only 9.9 per cent in the UK as a whole over the same period. If these comparative rates of growth continue into the 1990s, Wales would be one of the more prosperous regions of the UK - and in terms of socio-economic indicators fall within the 'south' rather than the 'north'. Unemployment in Wales, moreover, while still comparatively high, had decreased at a disproportionately rapid rate (1985-7) - from 14.9 to 13.0 per cent compared to a decline from 11.8 to 10.6 per cent in the UK overall.

A high level of investment from overseas is part of the reason for the improved performance of the Welsh economy. With only 5 per cent of the population of Great Britain, Wales attracted 21 per cent of Britain's inward investment in 1987 - more than any other region. Funds worth £158 million were scheduled to create 4,000 jobs - increasing the total numbers employed in the 300 foreign-owned firms in the principality to 50,000; or about 25 per cent of the manufacturing workforce. The USA was the largest source of inward investment (through such firms as Hoover, Monsanto and 3M), followed by Japan (in the case of Sony, Panasonic and Hitachi) and latterly West Germany (in respect of Bosch) (see Hetherington, 1987; Smith, 1988).

While the WDA - founded in 1976 by the Wilson government - has been instrumental in attracting foreign investment, it is committed to making indigenous enterprise the principal platform for the regeneration of the Welsh economy. To attract investment from both overseas and domestic sources, the WDA has invested funds in the reclamation of derelict land, developed several industrial parks, provided a range of fully serviced and often redeveloped sites, purpose-built premises and advisory services, in addition to financing the formation and growth of firms. By 1987, the WDA had built up an investment portfolio of £35 million in 400 companies and was the main source of venture capital in the principality (see Hetherington, 1987). Small-scale enterprise has been particularly important in the regeneration of the Welsh economy. Over the period of 1981-5, the number of self-employed increased by 27 per cent, and by 1986, 154,000 people were self-employed in Wales - only the South East and East Anglia having a greater number of firms of this kind (see Beresford, 1987).

Althouth the Welsh economy was recovering from its long period of decline through a process of industrial diversification, in 1988 Mr Walker launched a major attack on the problems of industrial decay in the South Wales valleys.

Table 5.4 Department of Trade and Industry expenditure in Wales, 1984–92

	1984–5*	1985–6*	1986–7*	1987–8*	1988–9†	1989–90‡	1990–1‡	1991–2‡
Total Department of Trade and Industry expenditure § (cash £m)	148.0	156.0	170.0	156.0	195.0	212.0	210.0	200.0
Increase from previous year, %	4.2	5.4	9.0	–8.2	25.0	8.7	–0.9	–4.8
Expenditure by Welsh Development Agency‖ (cash £m)								
Gross	70.0	63.5	72.0	98.6	113.3	130.4	113.0	130.0
Net	41.0	33.2	35.2	57.0	64.8	72.5	73.7	75.3
Increase from previous year in net expenditure, %	–7.4	–19.0	6.0	61.9	13.7	11.9	1.7	2.2

Notes

* Outturn.
† Estimated outturn.
‡ Planned.
§ Includes expenditure on regional selective assistance, regional development grants, enterprise-initiative consultancy schemes, grants to local education authorities to support career services, the Welsh Development Agency and the Development Board for Rural Wales.
‖ Gross expenditure includes net expenditure (the DTI allocation) *plus* receipts from the sale of land and property and from rents.

(*Source*: Treasury, *The Government's Expenditure Plans, 1989–90 to 1991–92, Wales*, Cmnd. 617.)

Under this initiative, £500 million were to be allocated over three years to improve economic, environmental and social conditions in the valleys and to lever £1 billion of private investment into a wide range of development projects. Unemployment was already falling in the valleys of Gwent, Mid Glamorgan and West Glamorgan (from 50,000 to 37,000, 1986-8), and it was hoped it would decline by a further 30,000 by 1991 -which in percentage terms would put the valleys on a par with the South East (see Hoyland and Tirbutt, 1988).

The purpose of this brief review of economic development in Wales is to show that a depressed region can begin to display signs of recovery if, first, it has a development agency that acts as a catalyst for regeneration; second, a senior minister is appointed with special responsibility for the region and with sufficient power to command - for the benefit of the region - an increased share of public expenditure; and, third, indigenous enterprise is nurtured and encouraged to perform a major role in development. Inward investment, to some extent, might also contribute to growth. However, despite the undoubted achievements of regional policy in Wales in the late 1980s, it might only be through the medium of elected regional councils (or assemblies) that it is possible to formulate and administer accountable policies on an appropriate scale to reduce substantially disparities between the depressed regions (including Wales) and the more prosperous 'south'.

Lessons from the South East

Whereas much of Wales has been intermittently depressed since the 1930s, the South East has enjoyed the benefits of continual growth for at least an equivalent period of time - the distinction between the two regional economies being clearly evident throughout much of the 1980s (Table 5.3). But the South East has also incurred the very substantial (external) costs of growth, particularly since the abolition of the South East Regional Economic Planning Council in 1979 and the Greater London Council in 1986 with their respective regional and strategic planning powers. In the 1980s, the strict preservation of the Green Belt appeared to be the over-riding planning goal of public policy in the South East - regardless of the adverse effects of this policy on economic development and social needs in the region, while the only regional body with any interest in planning - the South East Regional Planning Conference (SERPLAN) - was a loose-knit group of local authorities devoid of statutory power.

If positive economic and spatial planning were to re-emerge in the South East in the 1990s, its immediate responsibilities would be fourfold. First, there would be a need to eliminate the costs and other adverse effects of overheating; second, it would be necessary to stimulate further growth - not just to the advantage of the region but to the benefit of the economy of the UK overall; third, on egalitarian grounds - as well as for greater efficiency - it would be desirable to redistribute resources away from the more prosperous/fast-growing areas of the South East to the less affluent/comparatively depressed parts of the region; and, finally, policies should be designed to tackle effectively the increasing problems of multi-deprivation in inner London.

A major consequence of overheating in the region is an inadequate supply of transportation services, health services and housing in relation to demand or need: a deficiency that in absolute terms significantly exceeds that of any northern region. The public sector must, therefore, invest heavily in infrastructure, social and community facilities (particularly the National Health Service) and social housing throughout the region. With regard to transport in Greater London and the rest of the South East, it must be recognized that the volume of central government subsidy has for many years been considerably higher than elsewhere in Britain. On the basis of the principle that the provision of local services should be financed locally rather than nationally, Mr Michael Fallon, MP (1988), suggested that since the real beneficiaries of transport subsidies are the local economies where commuters live and spend most of their income, local councils – rather than the Exchequer – should subsidize commuting fares (out of the proceeds of local rates and/or the community charge). While it might be difficult to justify this principle in respect of other local services, locally derived subsidies (in the case of public transport) could provide a solution to the increasing problems of traffic congestion in London and the rest of the region, and would particularly ease the problems of accessibility in inner London to the benefit of local industry. Directly pricing the use of roads could provide a further revenue resource. Currently, the private motorist fails to cover his or her full cost of using roads – congestion costs in the capital incurring firms alone £15 billion per annum (according to the Confederation of British Industry, 1989). A means of pricing the use of urban roads through a system of zonal licensing, electronic vehicle-monitoring or electronic metering, would not only provide revenue for the maintenance and improvement of the existing road network, but (together with rate/community-charge revenue) would also facilitate a very high level of public transport subsidy that (together with road pricing) would shift large numbers of travellers from private cars to low-fare public bus and rail services (see Balchin, Bull and Kieve, 1988; Jacques, 1989). Clearly, within the South East, there would be the need for a high level of investment in a new regional rail network[3] – similar to the RER network in the Ile de France (Paris) region – where express services from the Outer South East would connect with a modernized London underground system (see Hall, 1989).

To ensure that the rate of growth experienced in the South East in the 1980s is maintained (or increased), full employment throughout the region will need to be secured in the 1990s – a difficult aim since a further loss of jobs in manufacturing, construction and transport services (not matched by the growth in service employment) seems probable, unemployment at late-1980s levels (or higher) might persist, a high natural increase in the working population is forecast, and a high rate of migration from poorer regions is anticipated (see Wood, 1987; Stott, 1988). It is probable, moreover, that there will be only a negligible rate of migration from the South East to the north (despite 'quality-of-life' considerations), since job prospects diminish northward and migrants would risk forfeiting their long-term competitive position in the housing market of the south (see Champion, 1989). To facilitate growth in the South East, it will be necessary to apply policies similar to some of those directed at the north (Lock,

1988). Apart from the government facilitating a high volume of public-sector investment in a modern infrastructure, it should induce the private sector (through the medium of a regional development agency or enterprise boards) to invest in, for example, ' "state of the art" technological infrastructure ... global communications and transport facilities [and] ... the development of sophisticated and attractive premises' (Lock, 1988, p. 88).

A redistribution of resources within the region, from overheated areas to areas of special need, would involve the investment of public-sector funds in development and urban regeneration in the more depressed parts of the Outer South East (to accommodate indigenous growth – including housing). As in the inner cities, public-sector investment could be expected to lever private funds into joint-development schemes as well as having broad multiplier effects locally.

Despite the application of a wide range of measures aimed at regenerating the economy of inner London over the last two (or more) decades, the possibilities of a further decline in manufacturing and service employment in the capital in the 1990s cannot be discounted. Where employment is generated, for example in financial services, there will be comparatively few job opportunities for unemployed labour in the inner-London boroughs. Under relatively free-market conditions, most employment growth will continue to be generated around the rest of the South East – decentralization being particularly marked during periods of reflation. As Elias (1982) suggested, only a policy directed at reducing industrial costs and stimulating demand would favour London. In practical terms, this would necessitate, for example, the introduction of policies aimed at reducing the cost of land acquisition (for development purposes) and the volume of road congestion; while demand from the business sector could be generated by facilitating the simultaneous expansion of interdependent industries and firms, and from the personal sector by ensuring that low-paid part-time work was transformed into higher-paid full-time employment, and that the capital can support at least its present population by providing an adequate supply of affordable housing and a greater choice of tenure.

It is important to recognize than an effective system of regional planning in the South East is not only essential for the economic well-being of the region but is also necessary for the realization of national macroeconomic objectives. While a more rational distribution of resources within the South East would stimulate growth and reduce the rate of inflation in the region, a higher level of prosperity in the South East (and a wider distribution of wealth and incomes within the region) would provide the revenue for the Exchequer to increase the volume of development aid to the north and so reduce the level of unemployment in the recipient areas. It is by means of this process of redistribution rather than through a misplaced attempt to level down the economy of the South East by restrictions on growth, that the north-south divide could best be narrowed. It is important, however, to recognize that without an effective fiscal mechanism for redistributing wealth and incomes to the north, prosperity is unlikely to trickle down significantly to the poorer regions of Britain, and that the continuing overheating of the South East would severely affect the macroeconomy overall by increasing the rate of inflation, by sucking in imports to satisfy a high level of consumer demand in the region and by necessitating an increasing volume of subsidies to finance infrastructure, mortgage interest relief and the rising cost of rentals on government properties (see Massey, 1988; Shepley, 1989).

In broad terms, the above review of policy priorities could equally apply to the other regions of the 'south' - such as East Anglia and the South West - although, except in the case of Bristol, there is an absence of an inner-city dimension. Essentially, the lessons to be learned from the South East are, first, that even in an area of economic growth there is a very great need for regional planning to tackle the problems of shortages, to stimulate growth and to redistribute resources (and if there are inner-city problems, to regenerate the inner-city economy). Second, as in the north, there is a need for elected regional councils to be set up to formulate and to apply a comprehensive range of accountable policies to achieve the objectives of economic and spatial planning discussed above. There would also be a need to replace the defunct Greater London Council with a powerful but 'slimline' metropolitan authority with considerable responsibilities for planning and transportation in the capital - but with fewer unwieldly powers of local government than hitherto (see Hall, 1989). Clearly, a new Greater London authority would need to harmonize its policies with both the regional council and the London boroughs. Third, it is essential that development agencies (akin to the SDA and WDA) are established to act as catalysts for growth. In addition, rather more than in the north, private-sector resources could be expected to perform a major role in development/regeneration. Private financial and industrial capital could thus be utilized on a significant scale by regional-development agencies and local-enterprise boards in each of the southern regions.

Lessons from Continental Europe

Whereas Great Britain is probably the most highly centralized, advanced capitalist State in the world - in terms of both government and private-sector industry - in the rest of the European Community power has either been devolved to the regions or has never been heavily concentrated in any one area.

In France in the 1980s, there has been a gradual transformation from non-elected regional authorities (or 'think tanks' of local and national politicians) to directly accountable councils. For example, the Ile de France has emerged as an elected regional planning authority (with powers to determine the overall framework of land use through the medium of a regional plan), as a development agency charged with the responsibility of promoting the region's advantages to potential employers, and with a budget allocated chiefly to the provision of the RER network of regional rail services. At a more local level, Economic Expansion Committees (funded by *département* councils) have attempted to bring together a cross-section of local expertise to establish a consensus approach to development and to help form networks of investors and professionals, companies and support services (see Marquand, 1987).

In West Germany, although there has been an absence of direct government involvement in industrial development at a regional (*lander*) level, legislation and regulations control the ways in which chambers of commerce (Industrie und HandelsKammern) help firms to satisfy their consultancy, information and financial needs and how *handwerk* associations (of smaller 'artisanal' firms) can utilize support networks of public- and private-sector educational, research, information, finance and marketing services. Marquand (1983) pointed out that in Italy, by contrast (chiefly because of the tight clustering of industries, a long craft-tradition and strong local communities), firms – without State legislation or regulation – have formed themselves into either local trade associations (for example, in Emilia-Romagna) or operate through the medium of *'impannatori'* (middleman) (as in Prato) *vis-à-vis* technology transfer or development, other information services and marketing.

As Marquand (1987) implied, it should have been a serious cause of concern in Britain in the 1980s that (unlike France) the government had not provided an institutional framework at a regional level to generate development, that (unlike West Germany) it had not legislated to make others provide it and that (unlike Italy) firms lacked a sense of community and were of 'indigenous potential' has clearly been developed and applied throughout much of the European Community in recent years, but in Britain it remains to be exploited.

Proposals for change

In Britain, it has become increasingly recognized that regional economic and spatial planning necessitates – on the grounds of efficiency and accountability – the creation of regional tier of government. As Luttrell (1987) pointed out, Britain has become 'the apotheosis of a unitary state', but since this has weakened the regions (north and south) and is of dubious benefit to the national economy, it is necessary to establish, as in other countries, an 'institutional framework [that] ... enables the regions to help themselves, to have the powers and resources to do so, and to become decision-making centres in their own-right' (*ibid.*). Luttrell (and in broadly similar terms, the TCPA, 1989) proposed that, first, directly elected councils should be set up in each of the eight regions of England and in Scotland and Wales, and that they would receive powers devolved to them from central government (and from the transference of responsibilities from, for example, regional health authorities and, by implication, from a renationalized water industry). County councils would become redundant since their responsibilities would be absorbed by either the district authorities or the new regional councils.

Second, central government would apply its interregional policy through a Ministry of Regional Development (or by an existing single ministry such as the Department of Trade and Industry). Despite the merits of having senior government ministers with wide-ranging responsibilities for development within each region, it is unlikely that ministers of Cabinet rank could be appointed to head every regional department in Britain (notwithstanding the need to retain Secretaries of the State at the Welsh and Scottish Offices), and unless there was a substantial increase in regional aid, some ministers would gain and others would lose in their bids for Treasury funding. A single ministry, however, could co-ordinate intra-regional policy in each English region in full collaboration with the regional council. Throughout England, the regional council would implement its regional strategy partly through newly established development agencies (while the Scottish and Welsh offices would continue to collaborate with their respective development agencies).

Third, regional councils would need to decide how to use their financial resources – derived from central-government grants and precepts on district authorities. Regional councils would need to finance both strategic and physical planning, housing and health-service provision, highways and public transport, further and possibly higher education, industrial training, and water management and sewerage. Ultimately, regional councils would have tax-raising powers of their own.

Finally, with regard to development projects, regional councils could form partnerships with the private sector, and there would also be a need for the proposed Ministry of Regional Development (or Department of Trade and Industry) – in liaison with the regional councils and their development agencies – to co-ordinate the national and regional strategies of major firms.

The Royal Town Planning Institute (RTPI) (1987) made broadly the same recommendations, arguing that the creation of a regional level of government in the 1990s is almost inevitable, but (instead of focusing on the need to establish a Ministry of Regional Development) it proposed the setting up of a National Regional Development Council, chaired by the Prime Minister and comprising ministers and regional-authority representatives. The RTPI also suggested the creation of regional development agencies in England – modelled on the SDA and WDA. The President of the RTPI, Mr Chris Shepley (1989, p. 13), proposed (in addition) that 'every government department and nationalized (or privatized) industry should be required to produce a regular "Regional Impact Analysis" ' in relation to their spending plans, that traditional regional policy (rather than being abandoned) 'should be modified, improved, targeted, and vastly increased', and that (as if to confirm the need for a reinvigorated regional policy) a Royal Commission on the distribution of population and economic activity should be set up to examine what has changed since the publication of the Barlow Report in 1940 'and to consider what remedial measures, if any, should now be taken in the national interest' (Shepley, p. 11). This proposal would clearly have merit if the government of the day was merely complacent about the need for a major reform in regional policy; but if, on the one hand, it was strongly wedded to free-market policy or, on the other hand, it was already committed to a change in regional policy broadly along the lines advocated above, a Royal Commission would hardly be relevant.

Of the English regions, the North has probably been the strongest advocate of regional government. In the mid-1980s, the Northern Development Company (NDC) was formed to co-ordinate the development initiatives of the region's local authorities, the Northern Trade Union Congress and the Confederation of British Industry. Modelled to some extent on the SDA and WDA, the NDC was authorized to take over the North of England Development Council and its

Whitehall-funded £650,000 annual budget. However, compared to the £100 million annual expenditure of the SDA, the NDC's total yearly budget of £2 million paled into insignificance. Clearly, only a full-blooded regional development agency would put the North on the same footing as Scotland and Wales, and it was with this in mind that a northern group of MPs published a parliamentary bill in 1988 advocating a Northern Development Agency (and a directly elected regional council) (see Osmond, 1988). There have also been proposals for regional development agencies emanating from other regions, such as the North West and Yorkshire & Humberside – two regions given dire forecasts of job losses and stagnant output over the period 1990–2000 (Cambridge Econometrics and Northern Ireland Economic Research Centre, 1988). Spanning these two regions, an embryonic development agency, TransPennine, has already been set up with private finance with the aim of promoting co-ordinated private- and public-sector development within central wedge of England, broadly along the M62 route linking Liverpool and Hull.

In conclusion, it would seem essential that regional councils, development agencies and local-enterprise boards attempt to establish, in collaboration with the private sector, a consensus approach to regional development and the decentralization of power; and to encourage private firms (including small enterprises) to develop and utilize support networks of educational, research, information, consultancy, financial and marketing services.

Political options

As though to deny the existence of a north-south divide, the spatial policy of the third Thatcher government was very largely concerned with the economic regeneration of the inner city – and particularly with private-sector initiative as the means of bringing this about. Within the Conservative Party, nevertheless, there were calls for a more positive stance on the regions. Reference has been made to Mr Michael Heseltine's proposals to abolish tax privileges which were overwhelmingly benefiting the south (Heseltine, 1987, 1988); and Mr Leon Brittan (1987b) urged the dispersal of central government and national agencies away from the South East, the consideration – by government – of every major policy decision in terms of its impact on regional imbalance and the setting up of English development agencies (modelled on the SDA and WDA) as focuses for regional development. It was evident, however, that none of these proposals would be implemented by government in the 1980s. Regional-development agencies (and regional councils), for example, had been opposed by central administrations for many years since devolved power would impinge upon the responsibilities of Whitehall.

The Social Democratic Party, in contrast, strongly believed that the drift towards regional inequality could not be reversed until Britain became a country with several regional centres. Throughout the north, large-scale investment in roads, rail transport and airports must, in the view of the SDP, become a priority to counteract the pull of the South East, and English development agencies should be established alongside a strengthened SDA and WDA (SDP, 1988).

Whereas the Labour Party in its 1987 general-election manifesto (Labour Party, 1987) retreated from an earlier commitment to establish regional councils in at least the North, the North West and Yorkshire & Humberside, and was equivocal in some of its positive intentions, in its policy review for the 1990s it proposed a comprehensive range of regional policies (Labour Party, 1989). Influenced by the widening north-south divide and the prospects of an uneven impact of the Channel tunnel in the 1990s, the Labour Party – in its review – proposed that the regional implications of every governmental decision would in the future need to be analysed. In the party's view, public expenditure, notably on high-quality research and further education, must be more evenly distributed. It would also be necessary to set up regional-development agencies in England and a network of regional investment banks – funded and co-ordinated by a new British Investment Bank – to offer long-term investment capital particularly to indigenous industry.

The Labour Party also proposed a considerable degree of devolution. Up to ten elected regional councils (each containing about 5 million people) would be set up in England, in part to exercise power decentralized from central government, but also to 'absorb, under democratic control, the functions exercised by non-elected boards and quangoes' (Labour Party, 1989, p. 57). In particular, the councils would take over the responsibilities of the regional health authorities, the water and sewerage undertakings, and the counties (in respect of strategic planning). Although the party, when in government, would set up an elected Welsh Assembly, at the time of the policy review it had not finalized its proposals concerning the assembly's responsibilities and funding, but there was a firm commitment to establish a Scottish Assembly with legislative and tax-raising powers and with responsibilities over the SDA, the Highlands and Islands Development Board and water.

In the 1990s, there is clearly a need to replace the atrophied and often-contradictory regional-planning policies of the previous decade. To compete effectively with other member states of the European Community after 1992, it is essential that economic efficiency is increased throughout the whole of the UK, and that both interregional and intra-regional disparities in prosperity are significantly narrowed or eliminated as soon as possible.

Notes

1. The Civil Service Commission (1989) reported that, in London and the Home Counties in 1988, only 60 per cent of the 3,658 executive officer vacancies and only 25 per cent of the 326 architect and 121 accountancy vacancies could be filled. Elsewhere in the UK, a considerably higher proportion of vacant jobs were filled – up to 90 per cent, for example, in the case of executive grade posts. The commission also reported that applications for civil-service appointments had dropped by 10 per cent following a decision to advise candidates for executive grade posts that they should not apply if they were not willing to work in London and the South East.

2. In May 1989, British Rail announced that it would spend £1.4 billion to expand capacity on congested Network SouthEast commuter lines by 30 per cent over the period to 1994. A further £200 million would be invested in the Heathrow Airport link (which will

carry up to 40,000 passengers a day from 1991) and £45 million would be spent on the Stansted Airport link (which will cater for 30,000 passengers per day). In addition, by the late 1990s, the Channel tunnel rail link will carry up to 16,000 commuters per hour at half the journey time of the 1980s. It is unlikely, however, that these largely *ad-hoc* developments will create a thoroughly modern rail system at least on a par with that of the Ile de France. An improved drive-and-ride facility (as is planned through the provision of more car-parking capacity at suburban stations) may be no substitute for a fully integrated rail and underground system.

3. In August 1989, the government announced that the Scottish Development Agency would be merged with the Training Commission in Scotland to create Scottish Enterprise.

4. The Highlands and Islands Development Board was given an added social dimension in 1989 and renamed Highland Enterprise.

BIBLIOGRAPHY AND REFERENCES

Alexander, I. (1979) *Office Location and Public Policy*, Longman, London.
Armstrong, H. and Riley, D. (1987) The 'north-south' controversy and Britain's regional problem, *Local Economy*, Vol. 2, pp. 93-105.
Armstrong, H. and Taylor, J. (1985) *Regional Economics and Policy*, Philip Allan, Oxford.
Armstrong, H. and Taylor, J. (1986) Regional policy: dead or alive?, *Economic Review*, Vol. 2, pp. 2-7.
Armstrong, H. and Taylor, J. (1987a) Job recovery must be nationwide, *Guardian*, 21 January.
Armstrong, H. and Taylor, J. (1987b) *Regional Policy: The Way Forward*, Employment Institute, London.
Ashcroft, R. K. and Taylor, J. (1979) The effect of regional policy on the movement of industry in Great Britain, in D. MacLennan and J. Parr (eds.) *Regional Policy: Past Experience and New Directions*, Martin Robertson, London, pp. 43-64.
Ashton, J. (1988) Life and health across the divide, *Town and Country Planning*, Vol. 57, pp. 276-78.
Association of London Authorities (1986) *A City Divided*, ALA, London.
Association of London Authorities (1988) *The Case for London in a Review of the European Commission's Structural Funds*, ALA, London.
Atkinson, R. (1988) *A Return to the Regions*, Bow Group, London.
Balchin, P. N. (1981) *Housing Improvement and Social Inequality*, Saxon House, Farnborough.
Balchin, P. N. (1989) *Housing Policy: An Introduction* (2nd edn), Routledge, London.
Balchin, P. N. and Bull, G. H. (1987) *Regional and Urban Economics*, Paul Chapman, London.
Balchin, P. N., Bull, G. H. and Kieve, J. L. (1988) *Urban Land Economics and Public Policy* (4th edn), Macmillan, London.
Ball, M. (1983) *Housing Policy and Economic Power*, Methuen, London.
Bank of England (1984) *Quarterly Bulletin*, March.
Barlow Report (1940) *Report of the Royal Commission on the Distribution of the Industrial Population*, Cmd. 6153, HMSO, London.
Bassett, K. and Short, J. (1980) *Housing and Residential Structure: Alternative Approaches*, Routledge & Kegan Paul, London.
Begg, I. G. and Cameron, G. C. (1988) High technology location and the urban areas of Britain, *Urban Studies*, Vol. 25, pp. 361-79.
Begg, I. G. and Moore, B. (1987) The changing economic role of Britain's cities, in V. A. Hausner (ed.) *Critical Issues in Urban Economic Development II*, Clarendon Press, Oxford, pp. 44-76.
Beresford, P. (1987) Golden future in the valleys, *The Sunday Times*, 29 April.

Black Committee (1980) Report of the working group on inequalities in health, in P. Townsend, N. Davidson and M. Whitehead (eds.) (1988) *Inequalities in Health: The Black Report and the Health Divide*, Penguin Books, Harmondsworth, pp. 44-226.

Black, D. and Liniecki, A. (1989) Regions demand high-speed links to channel tunnel, *The Independent*, 9 March.

Blazyca, G. (1983) *Planning is Good for You*, Pluto Press, London.

Boddy, M. (1987) Defence spending and the north/south divide, paper presented at the TCPA Annual Conference, Sheffield.

Bourne, L. S. (1981) *The Geography of Housing*, Edward Arnold, London.

Breheny, M., Cheshire, P. and Langridge, R. (1983) The anatomy of job creation? Industrial change in Britain's M4 corridor, *Built Environment*, Vol. 9, pp. 61-71.

Breheny, M., Hall, P. and Hart, D. (1987) *Northern Lights: A Development Agenda for the North in the 1990s*, Derrick, Wade & Walters, Preston and London.

Brittan, L. (1987a) Speech at Richmond, North Yorks., 7 August.

Brittan, L. (1987b) Speech at the TCPA Annual Conference, Sheffield, 30 August.

Brough, G. and Palmer, R. (1988) Northern green belt in danger, *The Sunday Times*, 20 November.

Buck, N., Gordon, I. and Young, K., with Ermisch, J. and Mills, L. (1986) *The London Employment Problem*, Oxford University Press.

Business Statistics Office (1981) *Business Monitor P A 1003. Report on the Census of Production, 1978. Analyses of United Kingdom Manufacturing (Local) Units by Employment Size*, HMSO, London.

Business Strategies Ltd (1989) *Regional Outlook*, BS Ltd, London.

Cambridge Econometrics and Northern Ireland Economic Research Centre (1988) *Regional Economic Prospects: Analysis and Forecasts to the Year 2000 for the Standard Regions of the United Kingdom*, Cambridge Econometrics (1985) Ltd, Cambridge.

Cameron, G. (1985) Regional economic planning - the end of the line, *Planning Outlook*, Vol. 28, pp. 8-31.

Carstairs, V. (1981) Small area analysis and health service research, *Community Medicine*, Vol. 3, pp. 131-9.

Centre for Local Economic Strategies (1989) *Channel Tunnel: Vicious Circle*, CLES, Manchester.

Champion, A. G. (1987) *Population Decentralization in Britain 1971-84*, Seminar Paper 49, Department of Geography, University of Newcastle upon Tyne.

Champion, A. G. (1989) Internal migration and the spatial distribution of population, in H. Joshi (ed.) *The Changing Population of Britain*, Blackwell, Oxford, pp. 110-32.

Champion, A. G. and Green, A. E. (1985) *The Booming Towns of Britain: The Geography of Economic Performance in the 1980s*, Discussion Paper 72, Centre of Urban and Regional Development Studies, University of Newcastle upon Tyne.

Champion, A. G. and Green, A. E. (1988) *Local Prosperity and the North South Divide: Winners and Losers in 1980s Britain*, Institute for Employment Research, University of Warwick.

Champion, A. G., Green, A. E. and Owen, D. W. (1988) House prices and local labour market performance: an analysis of building society data for 1985, *Area*, Vol. 20, pp. 253-63.

Champion, A. G., Green, A. E., Owen, D. W., Ellis, D. J. and Coombes, M. G. (1986) *Changing Places: Britain's Demographic, Economic and Social Complexion*, Edward Arnold, London.

Channel Tunnel Group (1985) *The Channel Tunnel Project: Employment and other Economic Implications*, CTG, London.

Channel Tunnel Joint Consultative Committee (1987) *A Strategy for Kent, Kent Impact Study: Second Report*, CTJCC, Maidstone, Kent.

Cheshire, P., Carbonaro, G. and Hay, D. (1986) Problems of urban decline and growth in EEC countries; or measuring degrees of elephantness, *Urban Studies*, Vol. 23, pp. 131-49.

Civil Service Commission (1989) *Annual Report 1988*, HMSO, London.
Clarke, H. (1986) Help small rural industries, paper presented at the TCPA Summer School, 14 September 1986, University of Nottingham.
Confederation of British Industry (1989) *Capital at Risk*, CBI, London.
Cooke, P. N. (1987) Britain's new spatial paradigm: technology, locality and society in transition, *Environment and Planning A*, Vol. 19, pp. 1289–301.
Cross, D. (1988) Counterurbanization in England and Wales: migration - the evidence, paper presented at the IBG Annual Conference, University of Loughborough.
Damesick, P. J. (1982) Strategic choice and uncertainty: regional planning in South East England, in R. Hudson and J. Lewis (eds.) *Regional Planning in Europe, London Papers in Regional Science*, Vol. 11, Pion, London.
Damesick, P. J. (1987) Regional economic change since the 1960s, in P. J. Damesick and P. A. Wood (eds.) *Regional Problems, Problem Regions and Public Policy in the United Kingdom*, Clarendon Press, Oxford, pp. 19–41.
Damesick, P. J. and Wood, P. A. (eds.) (1987) *Regional Problems, Problem Regions and Public Policy in the United Kingdom*, Clarendon Press, Oxford.
Danson, M. W., Lever, W. F. and Malcolm, J. F. (1980) The inner city employment problem in Great Britain: a shift-share approach, *Urban Studies*, Vol. 12, pp. 193–210.
Davies, P. (1989) Let them eat crumbs, *Guardian*, 16 May.
Dean, M. (1985) Budget to pump in inner city aid and cash, *Guardian*, 3 October.
Dennis, R. D. (1978) The decline of manufacturing employment in Greater London: 1966–74, *Urban Studies*, Vol. 15, pp. 63–73.
Dennis, R. D. (1980) The decline of manufacturing employment in London 1966–74, in A. Evans and D. Eversley (eds.) *The Inner City*, Heinemann, London, pp. 45–64.
Dennis, R. D. (1981) *Changes in Manufacturing Employment in the South East Region between 1976 and 1980*, Department of Trade and Industry, London.
Department of Employment (1985) *New Earnings Survey*, HMSO, London.
Department of Employment (1986) *New Earnings Survey*, HMSO, London.
Department of Employment (1987a) *Census of Employment*, HMSO, London.
Department of Employment (1987b) *New Earnings Survey*, HMSO, London.
Department of the Environment (1977a) *Inner London - Policies for Dispersal and Balance, Lambeth Inner Area Study*, HMSO, London.
Department of the Environment (1977b) *Policy for the Inner Cities*, Cmnd. 6845, HMSO, London.
Department of the Environment (1988) *English House Condition Survey, 1987*, HMSO, London.
Department of the Environment (1989) *The Future of Development Plans*, Cmnd. 569, HMSO, London.
Department of Trade and Industry (1981) *Industrial Movement in the United Kingdom, 1966–1975*, HMSO, London.
Department of Trade and Industry (1983) *Regional Industrial Development*, Cmnd. 9111, HMSO, London.
Department of Trade and Industry (1984) *Memorandum to SERPLAN*, DTI, London.
Department of Trade and Industry (1985) *The Balance of Trade in Manufactures*, Cmnd. 9697, HMSO, London.
Department of Trade and Industry (1986) *UK Regional Development Programme 1986–90* (17 volumes), Submission to the European Regional Development Fund, DTI, London.
Department of Trade and Industry (1988a) *DTI - The Department for Enterprise*, Cmnd. 278, HMSO, London.
Department of Trade and Industry (1988b) *Statement to the Public Inquiry into Proposals for a New Settlement at Foxley Wood*, DTI, London.
Department of Transport (1989) *Roads for Prosperity*, Cmnd. 693, HMSO, London.
Drewett, R. (1973) The developers: decision processes, in P. Hall, H. Grady, R. Drewett and R. Thomas (eds.) *The Containment of Urban England*, Vol. 2, Allen & Unwin, London, pp. 163–94.

Dunn, R., Forrest, R. and Murie, A. (1987) The geography of council house sales in England, *Urban Studies*, Vol. 24, pp. 47–59.

The Economist (1985) Green grows in the South East, 18 May.

The Economist (1988) Whitehall: moving out, 27 August.

Eddison, T. (1988) Too few roofs over too many heads, *Town and Country Planning*, Vol. 57, pp. 66–68.

Elias, P. (1982) Regional impact of national economic policies: a multi-regional simulation approach for the United Kingdom, *Regional Studies*, Vol. 16, pp. 335–44.

Elias, P. and Keogh, G. (1982) Industrial decline and unemployment in the inner city areas of Great Britain: a review of the evidence, *Urban Studies*, Vol. 19, pp. 1–15.

Elliott, L. (1988) Biting retail winds chill the south, *Guardian*, 17 December.

Evans, A. W. (1987) *House Prices and Land Prices in the South East – A Review*, House Builders Federation, London.

Fallon, M. (1988) Southern comfort, *Guardian*, 11 April.

Ferriman, A. (1989) NHS waiting lists hit record levels, *The Observer*, 15 January.

The Financial Times (1987a) Jobs divide brought sharply into focus, 8 January.

The Financial Times (1987b) Regional developments in Britain: the gap widens, 20 January.

The Financial Times (1988) Policy seeks to help regions develop their own potential, 13 January.

Fleming, M. and Nellis, J. (1989) *The Impact of the Community Charge*, Nationwide Anglia Building Society, London.

Fothergill, S. (1988) Geography of jobs in Mrs Thatcher's Britain, paper presented at the RTPI Summer School, St Andrews University.

Fothergill, S. and Gudgin, G. (1979) Regional employment change: a sub-regional explanation, *Progress in Planning*, Vol. 12, pp. 155–219.

Fothergill, S. and Gudgin, G. (1982) *Unequal Growth: Urban and Regional Employment Change in the United Kingdom*, Heinemann, London.

Fothergill, S., Kitson, M. and Monk, S. (1985) *Urban Industrial Change: The Causes of the Urban-Rural Contrast in Manufacturing Employment Trends*, Department of the Environment/Department of Trade and Industry, HMSO, London.

Fothergill, S., Kitson, M. and Monk, S. (1987) *Property and Industrial Development*, Hutchinson, London.

Fothergill, S. and Vincent, J. (1985) *The State of the Nation*, Pan, London.

Fox, A. J., Jones, D. R. and Goldblatt, P. O. (1984) Approach to studying the effect of socio-economic circumstances on geographical differences in mortality in England and Wales, *British Medical Bulletin*, Vol. 40, pp. 309–14.

Friends of the Earth (1989) *Action for People*, FOTE, London.

Frost, M. E. and Spence, N. A. (1983) Unemployment change, in J. B. Goddard and A. G. Champion (eds.) *The Urban and Regional Transformation in Britain*, Methuen, London, pp. 239–59.

Gardner, M. J., Winter, P. D., Taylor, C. P. and Acheson, E. D. (1983) *Atlas of Cancer Mortality in England and Wales*, Wiley, Chichester.

Gardner, M. J., Winter, P. D. and Barker, D. J. P. (1984) *Atlas of Mortality from Selected Diseases in England and Wales, 1968–78*, Wiley, Chichester.

Goddard, J. B. (1985) The impact of new technology on urban and regional structure in Europe, paper presented at the BAAS Annual Conference, University of Strathclyde.

Goddard, J. B. and Coombes, M. (1987) The north-south divide: some local perspectives, paper presented at the IEA Conference on the North-South Divide.

Gossop, C. (1989) Can the promise be fulfilled?, *Town and Country Planning*, Vol. 58, pp. 4–5.

Gould, A. and Keeble, D. (1983) New firms and rural industrialization in East Anglia, *Regional Studies*, Vol. 18, pp. 189–201.

Greater London Council (1985a) *LIS – London Industrial Strategy: Introduction*, GLC, London.

Greater London Council (1985b) *The Greater London House Condition Survey*, GLC, London.

Gripaios, P. (1977) The closure of firms in the inner city: the south east London case 1970-1975, *Regional Studies*, Vol. 11, pp. 1-6.

Guardian (1988a) Ignoring the regions, 12 January.

Guardian (1988b) Jobs growth 'is all in the south', 14 July.

Gudgin, G. and Fothergill, S. (1984) Geographical variations in the rate of formation of new manufacturing firms, *Regional Studies*, Vol. 18, pp. 203-6.

Gudgin, G., Moore, R. and Rhodes, J. (1983) The great divide, *The Sunday Times*, 16 January.

Gudgin, G. and Schofield, A. (1987) The emergence of the north-south divide and its projected future, paper presented at the TCPA Annual Conference, Sheffield.

Hall, P. (1962) *The Industries of London since 1981*, Hutchinson, London.

Hall, P. (1981) The geography of the fifth Kondratieff Cycle, *New Society*, 26 March.

Hall, P. (1985) Land of green ginger groups, *The Times*, 24 October.

Hall, P. (1987a) Flight to the green, *New Society*, 9 January.

Hall, P. (1987b) The anatomy of job creation: nations, regions and cities in the 1960s and 1970s, *Regional Studies*, Vol. 21, pp. 95-106.

Hall, P. (1989) *London 2001*, Unwin Hyman, London.

Hall, P., Breheny, M., McQuaid, R. and Hart, D. (1987) *Western Sunrise: The Genesis and Growth of Britain's Major High-Tech Corridor*, Allen & Unwin, London.

Hall, P., Gracey, H., Drewett, R. and Thomas, R. (1973) *The Containment of Urban England*, 2 vols., Allen & Unwin, London.

Halsall, M. (1987) Headhunters struggle to cope with regional gap, *Guardian*, 27 May.

Halsall, M. (1988) Survey attacks regional policy, *The Independent*, 13 June.

Hamnett, C. (1983) Regional variations in house prices and house price inflation, 1969-81, *Area*, Vol. 15, pp. 97-109.

Hamnett, C. (1988a) Housing the new rich, *New Society*, 22 April.

Hamnett, C. (1988b) Regional variations in house prices and house price inflation in Britain, 1969-1988, *The Royal Bank of Scotland Review*, September, pp, 29-40.

Hancock, T. and Dahl, L. J. (1986) *Healthy Cities: Promoting Health in the Urban Context*, Background paper for the WHO Healthy Cities Project, WHO, Copenhagen.

Hannah, I. and Kay, J. A. (1977) *Concentrations in Modern Industry*, Macmillan, London.

Hardman Report (1973) *The Dispersal of Government Work from London*, Cmnd. 5363, HMSO, London.

Harloe, M., Issacharoff, R. and Minns, R. (1974) *The Organization of Housing: Public and Private Enterprise in London*, Heinemann, London.

Harrison, M. and Brown, C. (1988) Senior Tories join critics of aid reforms, *The Independent*, 13 January.

Haskins, C. (1987) Material to bridge the great divide, *Guardian*, 24 March.

Haynes, R. (1988) The urban distribution of lung cancer mortality in England and Wales, 1980-1983, *Urban Studies*, Vol. 25, pp. 497-506.

Health Education Council (1987) *The Health Divide: Inequalities in Health in the 1980s*, HEC, London.

Henry, J. D. (1987) The effect of takeovers on the Scottish economy, paper presented at the TCPA Annual Conference, Sheffield.

Heseltine, M. (1987) The chance to compete, *Guardian*, 6 April.

Heseltine, M. (1988) Speech at Blackpool, 11 April.

Hetherington, P. (1987) Walker flourishes in political exile, *Guardian*, 28 October.

Hillier Parker Research (1982) *Industrial Contour Map, Rents and Yields*, Hillier Parker May & Rowden, London.

Hillier Parker Research (1987) *Industrial Rent and Yield Contours*, Hillier Parker May & Rowden, London.

Hogg, S. (1987) A nation of inheritors, *The Independent*, 30 November.

190	*Regional Policy in Britain*

House of Commons (1987) Government economic policy, HOC parliamentary debates,
Hansard, Vol. 108, no. 34, pp. 764-836, HMSO, London.
House of Commons Expenditure Committee (1975) *New Towns*, report, HMSO, London.
House of Lords Select Committee on Overseas Trade (1985) *Report*, Cmnd. 238-1,
HMSO, London.
Howe, G. M. (1986) Does it matter where I live?, *Transactions of the Institute of British*
Geographers, New Series, Vol. 11, pp. 387-411.
Howells, J. R. L. (1984) The location of research and development: some observations and
evidence from Britain, *Regional Studies*, Vol. 14, pp. 13-29.
Hoyland, P. and Tirbutt, S. (1988) Walker spells out valleys revival plan, *Guardian*, 15
June.
Hudson, R. and Williams, A. (1986) *The United Kingdom*, Paul Chapman, London.
Hughes, C. (1988a) Tories begin pressure to abolish shire counties, *The Independent*, 10
March.
Hughes, C. (1988b) Gould urges devolution to balance the regions, *The Independent*, 2
July.
The Independent (1987a) Two-nations job shock revealed, 7 January.
The Independent (1987b) Labour aims £6 billion jobs plan at the north, 9 January.
Jacques, M. (1989) Slow crawl on the road to prosperity, *The Sunday Times*, 9 April.
Jaskowiak, C. (1989) Shops begin to feel the pinch, *The Sunday Times*, 1 January.
Johnston, R. J. (1987) A note on housing tenure and voting in Britain, *Housing Studies*,
Vol. 2, pp. 112-21.
Johnston, R. J., Pattie, C. J. and Allsopp, J. G. (1988) *A Nation Dividing?*, Longman,
London.
Keeble, D. (1968) Industrial decentralization and the metropolis: the north west London
case, *Transactions of the British Institute of Geographers*, Vol. 44, pp. 1-54.
Keeble, D. (1976) *Industrial Location and Planning in the United Kingdom*, Methuen,
London.
Keeble, D. (1977) Spatial policy in Britain: regional or urban?, *Area*, Vol. 9, pp. 3-8.
Keeble, D. (1980a) The south east, in G. Manners *et al.* (eds.) *Regional Development in*
Britain, Wiley, Chichester, pp. 101-75.
Keeble, D. (1980b) Industrial decline, regional policy and the urban-rural manufacturing
shift in the United Kingdom, *Environment and Planning A*, Vol. 12, pp. 945-62.
Keeble, D. (1986) The changing spatial structure of economic activity and metropolitan
decline in the United Kingdom, in H. J. Ewars, H. Matzerath and J. B. Goddard (eds.)
The Future of the Metropolis, de Gruyter, Berlin.
Keeble, D. (1987) Industrial change in the United Kingdom, in W. F. Lever (ed.) *Industrial*
Change in the United Kingdom, Longman, London, pp. 1-20.
Keeble, D. and Gould, A. (1986) Entrepreneurship and manufacturing firm formation in
rural regions: the East Anglian case, in M. J. Healey and B. W. Ilbery (eds.)
Industrialization of the Countryside, Geobooks, Norwich, pp. 197-220.
Keeble, D. and Kelly, T. (1986) New firms and high technology in the United Kingdom:
the case of computer electronics, in D. Keeble and E. Wever (eds.) *New Firms and*
Regional Development in Europe, Croom Helm, Beckenham.
Killick, T. (1983) Manufacturing plant openings, 1976-80, *British Business*, Vol. 17, pp.
466-8.
Labour Party (1987) *Election Manifesto*, LP, London.
Labour Party (1989) *Meet the Challenge. Make the Change: A New Agenda for Britain*,
LP, London.
Lee, C. H. (1986) *The British Economy since 1700: A Macroeconomic Perspective*,
Cambridge University Press.
Lever, W. F. (1982) Urban scale as a determinant of employment growth or decline, in L.
Collins (ed.) *Industrial Decline and Regeneration*, Department of Geography,
University of Edinburgh.

Lewis, J. (1988) Walker fights to bring prosperity to the scarred Welsh valleys, *Guardian*, 30 September.

Lock, D. (1988) South east boom - Yorkshire gloom, *Town and Country Planning*, Vol. 57, pp. 84–5.

Lock, D. (1989) *Riding the Tiger - Planning the South of England*, Town and Country Planning Association, London.

London Research Centre (1988) *Access to Housing in London. A Report based on the Results of the London Housing Survey 1986-7*, LRC, London.

Luttrell, W. F. (1987) A programme for regional development: what could be done in five years, paper presented at the TCPA Annual Conference, Sheffield.

MacInnes, J. (1988) *The North-South Divide: Regional Employment Change in Britain, 1975-87*, Discussion Paper 34, Centre for Urban and Regional Research, University of Glasgow.

Maclennan, D. and Parr, J. B. (eds.) (1979) *Regional Policy: Past Experiences and New Directions*, Martin Robertson, London.

Macmillan, H. (1938) *The Middle Way: A Study of the Problem of Economic and Social Progress in a Free and Democratic Society*, Macmillan, London.

McCormick, B. (1983) Housing and unemployment in Britain, *Oxford Economic Papers*, Vol. 35.

McCrone, G. (1969) *Regional Policy in Britain*, Allen & Unwin, London.

McGhie, C. (1988) Soaring land prices set to fuel prosperity boom, *The Sunday Times*, 20 November.

Manners, G. (1976) Reinterpreting the regional problem, *Three Banks Review*, September, pp. 35–50.

Manpower Services Commission (1987) *London Labour Market Report*, MSC, London.

Marquand, J. M. (1980) *Measuring the Effects and Costs of Regional Incentives*, Working Paper 32, Government Economic Service, Civil Service College, London.

Marquand, J. M. (1983) The changing distribution of service employment, in J. B. Goddard and A. G. Champion (eds.) *The Urban and Regional Transformation of Britain*, Methuen, London, pp. 99–134.

Marquand, J. M. (1987) The economic benefits of regional government, paper presented at the TCPA Annual Conference, Sheffield.

Marsh, P. (1983) Britain's high technology entrepreneurs, *New Statesman*, 10 November.

Martin, R. L. (1982) Job loss and the regional incidence of redundancies in the current recession, *Cambridge Journal of Economics*, Vol. 6, pp. 375–96.

Martin, R. L. (1986) Thatcherism and Britain's industrial landscape, in R. L. Martin and R. E. Rowthorne (eds.) *The Geography of Deindustrialization*, Macmillan, London, pp. 238–90.

Martin, R. L. (1987) Mrs Thatcher's Britain: a tale of two nations, *Environment and Planning*, Vol. 19.

Martin, R. L. (1988) The political economy of Britain's north-south divide, *Transactions of the Institute of British Geographers*, New Series, Vol. 13, pp. 389–418.

Martin, R. L. and Rowthorn, R. E. (eds.) (1986) *The Geography of Deindustrialization*, Macmillan, London.

Mason, C. M. (1987) Venture capital in the United Kingdom: a geographical perspective, *National Westminster Bank Review*, May, pp. 47–59.

Mason, C. M. and Harrison, R. T. (1989) Small firms policy and the 'north-south' divide in the United Kingdom: the case of the Business Expansion Scheme, *Transactions of the Institute of British Geographers*, New Series, Vol. 14, pp. 37–58.

Massey, D. (1984) *Spatial Divisions of Labour*, Macmillan, London.

Massey, D. (1988) A new class of geography, *Marxism Today*, May.

Massey, D. and Meegan, R. A. (1978) Industrial restructuring versus the cities, *Urban Studies*, Vol. 15, pp. 273–88.

Merrett, S. (1976) Gentrification, in *Housing Class in Britain*, Political Economy of Housing Workshop, London, p. 44-9.

Merrett, S. with Gray, F. (1982) *Owner Occupation in Britain*, Routledge & Kegan Paul, London.

Metcalf, D. and Richardson, R. (1976) Unemployment in London, in G. D. N. Worswick (ed.) *The Concept and Measurement of Involuntary Unemployment*, Allen & Unwin, London.

Ministry of Labour (1944) *White Paper on Employment Policy*, Cmd. 6527, HMSO, London.

Mintel Publications Ltd (1988) *Regional Lifestyle*, Mintel, London, £550.

Moore, B. and Rhodes, J. (1973) Evaluating the effects of British regional economic policy, *Economic Journal*, Vol. 83, pp. 87–110.

Moore, B. and Rhodes, J. (1976) A quantitative assessment of the effects of the regional employment premium and other regional policy instruments, in A. Whiting (ed.) *The Economics of Industrial Subsidies*, HMSO, London.

Moore, B., Rhodes, J. and Tyler, P. (1981) *The Impact of Regional Policy on Regional Labour Markets*, Department of Applied Economics, University of Cambridge.

Moore, B., Rhodes, J. and Tyler, P. (1984) *Geographical Variations in Industrial Costs*, Discussion Paper 12, Department of Land Economy, University of Cambridge.

Moore, B., Rhodes, J. and Tyler, P. (1986) *The Effects of Government Regional Economic Policy*, Department of Trade and Industry, HMSO, London.

Muellbauer, J. (1986) How house prices fuel wage rises, *The Financial Times*, 23 October.

Muellbauer, J., Bover, O. and Murphy, A. (1988) *Housing, Wages and the UK Labour Market*, Discussion Paper 268, Centre for Economic Policy Research, London.

Muellbauer, J. and Murphy, A. (1988) *UK House Prices and Migration: Economic and Investment Implications*, Shearson Lehman Hutton, London.

National Federation of Housing Associations (1985) *Inquiry into British Housing: Report*, NFHA, London.

Nationwide Anglia Building Society (1989) *House Prices in 1988*, NABS, London.

Naughtie, J. (1986) Catalogue of decay reveals jobless will stay above 3 million, *Guardian*, 20 October.

New Society (1987) The north/south jobs divide, 9 January.

Nicholson, G. (1989) A model of how not to regenerate an urban area, *Town and Country Planning*, Vol. 58, pp. 52–55.

North of England Regional Consortium (1988) *State of the Region Survey*, NERC, Leeds.

O'Farrell, P. N. (1985) Manufacturing employment change and establishment size, *Area*, Vol. 17, pp. 35–43.

Oakley, R., Rothwell, R. and Cooper, S. (1989) *Management of Innovation in High Technology Small Firms*, Pinter Publishers, London.

Organ, J. D. (1987) Retail and residential competition hike industrial land prices, *Estates Times Supplement*, March.

Osmond, J. (1988) *The Divided Kingdom*, Constable, London.

Owen, D. H., Coombes, M. G. and Gillespie, A. E. (1983) *The Differential Performance of Urban and Rural Areas in the Recession*, Discussion Paper 49, Centre of Urban and Regional Development Studies, University of Newcastle upon Tyne.

Parsons, D. W. (1986) *The Political Economy of Regional Policy*, Croom Helm, Beckenham.

Pond, C. (1988) Letter to *The Independent*, 11 January.

Prais, S. J. (1976) *The Evolution of Giant Firms in Britain*, Cambridge University Press.

Randall, J. N. (1987) Scotland, in P. J. Damesick and P. A. Wood (eds.) *Regional Problems, Problem Regions and Public Policy in the United Kingdom*, Clarendon Press, Oxford, pp. 218–37.

Regional Studies Association (1983) *Report of an Inquiry into Regional Problems in the United Kingdom*, Geobooks, Norwich.

Reward Group (1989) *Cost of Living: Regional Comparisons*, RG, Diamond Way, Stone, Staffs, April.

Ridley, N. (1987) Speaking on 'The London Programme', London Weekend Television, 29 May.

Roberts, J. L. and Graveling, P. A. (eds.) (1985) *The Big Kill: Smoking Epidemic in England and Wales*, published for the Health Education Council and the British Medical Association by the North West Regional Health Authority.

Robson, B. (1987) Introducing the report of the Town and Country Planning Association inquiry into the north-south divide, TCPA Annual Conference, Sheffield.

Robson, B. (1988) Taxing the divide, *Town and Country Planning*, Vol. 57, pp. 130-1.

Roger Tym and Partners (1984) *Monitoring Enterprise Zones: Year Three Report*, Deparment of the Environment, London.

Rogerson, R., Findlay, A. and Morris, A. (1988) *A Report on Quality of Life in British Cities*, Department of Geography, University of Glasgow.

Rogerson, R., Morris, A., Findlay, A., Paddison, R. and Henderson, J. A. (1989) *Intermediate-Sized Cities in Britain: A Comparative Study of Quality of Life*, Department of Geography, University of Glasgow.

Royal Town Planning Institute (1987) *Strategic Planning for Regional Potential*, RTPI, London.

Segal Quince and Partners (1985) *The Cambridge Phenomenon: The Growth of High Technology in a University Town*, SQP, Cambridge.

Shepley, C. (1989) South Easternization. Fact or Fantasy? Benefit or burden?, *The Planner*, Vol. 75, pp. 11-13.

Smallwood, C. (1988) Thatcherisms's next big move, *The Sunday Times*, 31 January.

Smith, I. (1988) Where there's a welcome for the world, *The Observer*, 22 February.

Smith, M. and Travis, A. (1988) Young unveils his Department of Enterprise, *Guardian*, 13 January.

Social Democratic Party (1988) *Turning the Tide of Decline in the Regions*, SDP, London.

Southall, H. (1988) The origins of the depressed areas: unemployment, growth and regional structure in Britain before 1914, *Economic History Review*, Vol. 41, pp. 236-58.

Steer Davis and Gleave (1989) *The Right Track into Europe*, Transport 2000, London.

Stott, M. (1988) Not forgetting the south-south divide, *Town and Country Planning*, Vol. 57, pp. 237-39.

The Sunday Times (1987a) The nonsense of north-south, 11 January.

The Sunday Times (1987b) Two nations – the false frontier, 11 January.

Thunhurst, C. (1985) *Poverty and Health in the City of Sheffield*, Environmental Health Department, Sheffield City Council.

Town and Country Planning Association (1987) *North-South Divide - A New Deal for Britain's Regions*, TCPA, London.

Town and Country Planning Association (1989) *Bridging the North-South Divide*, TCPA, London.

Townsend, A. R. (1977) The relationship of inner city problems to regional policy, *Regional Studies*, Vol. 11, pp. 225-51.

Townsend, A. R. (1983) *The Impact of Recession on Industry, Employment and the Regions, 1976-81*, Croom Helm, Beckenham.

Townsend, A. R. (1986) The location of employment growth after 1978: the surprising significance of dispersed centres, *Environment and Planning A*, Vol. 18, pp. 529-45.

Townsend, A. R. (1987) Regional policy, in W. F. Lever (ed.) *Industrial Change in the United Kingdom*, Longman, London, pp. 223-39.

Townsend, A. R. and Peck, F. W. (1985) The geography of mass redundancy in named corporations, in M. Pacionne (ed.) *Progress in Industrial Geography*, Croom Helm, Beckenham, pp. 174-218.

Tyler, P. (1987) The success of traditional regional policy, paper presented at the TCPA Annual Conference, Sheffield.

Tyler, P. and Rhodes, J. (1986) *South-East Employment and Housing Study*, Discussion Paper 15, Department of Land Economy, University of Cambridge.

Tyler, P., Moore, B. and Rhodes, J. (1988) *Geographical Variations in Costs and Productivity*, Department of Trade and Industry, London.

Vickerman, R. W. (1987) The Channel tunnel: consequences for regional growth and

194 *Regional Policy in Britain*

development, *Regional Studies*, Vol. 21, pp. 187-97.

Watts, D. (1988) Administrative head offices: a northern perspective, paper presented at the IBG Annual Conference, Loughborough.

Webber, R. and Craig, J. (1976) Which local authorities are alike?, *Population Trends*, Vol. 5, pp. 13-19.

Whitehead, M. (1988) The health divide, in P. Townsend, N. Davidson and M. Whitehead (eds.) *Inequalities in Health: The Black Report and the Health Divide*, Penguin, Harmondsworth.

Whittington, R. C. (1983) Regional bias in new firm foundation in the United Kingdom, *Regional Studies*, Vol. 18, pp. 253-6.

Williams, I. (1988) Midlands back in business, *The Sunday Times*, 20 November.

Williams, I. (1989) Mortgages stifle south's holidays, *The Sunday Times*, 1 January.

Wilsher, P. and Cassidy, J. (1987) Two nations: the false frontier, *The Sunday Times*, 11 January.

Wintour, P. (1988) South-east 'given' 60 pc of tax cuts, *Guardian*, 4 August.

Wood, P. A. (1987) The south east, in P. J. Damesick and P. A. Wood (eds.) *Regional Problems, Problem Regions and Public Policy in the United Kingdom*, Clarendon Press, Oxford, pp. 64-94.

Wray, I. (1987) It's time to build a new policy for the regions, *Guardian*, 21 January.

Young, H. (1989) *One of Us*, Macmillan, London.

AUTHOR INDEX

Acheson, E.D. 144, 145
Alexander, I. 66
Allsopp, J.G. 39, 60, 62
Armstrong, H. 10, 25, 26, 64, 65, 70, 74, 76, 77, 161, 162, 165
Ashcroft, R.K. 71
Ashton, J. 145
Association of London Authorities 102, 107, 108, 111, 120, 122, 147, 149
Atkinson, R. 66, 71

Balchin, P.N. vii, 65, 85, 89, 166, 167, 170, 177
Ball, M. 139
Bank of England 42
Barker, D.J.P. 51, 54, 144, 145
Barlow Report 64, 65
Bassett, K. 139
Begg, I.G. 12, 13, 15
Beresford, P. 174
Black Committee 50
Black, D. 22
Blazyca, G. 170
Boddy, M. 80, 163, 165
Bourne, L.S. 39
Bover, A. 47
Breheny, M. 8, 9, 13, 15, 99, 101, 129, 130, 134, 151, 164, 170
Brittan, L. 95, 182
Brough, G. 105
Brown, C. 77
Buck, N. 85, 108, 109, 123, 136
Bull, G.H. viii, 65, 89, 166, 167, 170, 177
Business Statistics Office 103
Business Strategies Ltd. 26

Cambridge Econometrics and Northern Ireland Economic Research Centre 1, 182

Cameron, G.C. 13, 15, 103
Carbonaro, G. 82
Carstairs, V. 145
Cassidy, J. 102, 150, 153
Centre for Local Economic Strategies 22
Champion, A.G. 22, 41, 56, 57, 129, 139, 147, 151, 177
Channel Tunnel Group 20
Channel Tunnel Joint Consultative Committee 22
Cheshire, P. 9, 82
Civil Service Commission 182
Clarke, H. 133
Confederation of British Industry 109, 179
Cooke, P.N. 12
Coombes, M.G. 12, 22, 45, 76
Cooper, S. 163
Craig, J. 129
Cross, D. 133

Dahl, L.J. 145
Damesick, P.J. 12, 18, 72, 74, 94, 95, 105, 108, 109, 110, 171
Danson, M.W. 109
Davidson, M. 50
Davies, P. 81
Dean, M. 94
Dennis, R.D. 72, 109, 136
Department of Employment 10, 26, 106, 149
Department of the Environment 28, 85, 105
Department of Trade and Industry 7, 10, 26, 64, 72, 76, 160
Department of Transport 164
Drewett, R. 139
Dunn, R. 47, 62, 143

Economist 142, 163, 165
Eddison, T. 143
Elias, P. 109, 178
Elliott, L. 126
Ellis, D.J. 22
Ermisch, J. 85, 108, 109, 123, 136
Evans, A.W. 139

Fallon, M. 177
Ferriman, A. 146
Findlay, A. 153
Financial Times 9, 77
Fleming, M. 166
Forrest, R. 47, 62, 143
Fothergill, S. 5, 13, 17, 20, 25, 67, 72,
    87, 103, 109, 110, 136, 158, 160
Fox, A.J. 53, 146
Friends of the Earth 94
Frost, M.E. 26

Gardner, M.J. 51, 54, 144, 145
Gillespie, A. E. 12, 76
Goddard, J.B. 18, 46
Goldblatt, P.O. 54, 146
Gordon, I. 85, 108, 109, 123, 136
Gossop, C. 164
Gould, A. 9, 20
Graveling, P.A. 54
Greater London Council 98, 107, 120,
    143, 150
Gray, F. 139
Green, A.E. 22, 41, 56, 139, 147, 151
Gripaios, P. 109
Guardian 77
Gudgin, G. 8, 20, 25, 36, 45, 67, 72, 87,
    98, 102, 103, 109, 126, 136, 159, 160

Hall, P. 8, 9, 13, 15, 72, 99, 101, 109,
    129, 130, 133, 134, 136, 142, 151, 164,
    170, 177, 179
Halsall, M. 161, 162
Hamnett, C. 39, 45, 137
Hancock, T. 145
Hannah, I. 109
Harloe, M. 139
Harrison, M. 77
Harrison, R.T. 18, 79
Hart, D. 8, 13, 15, 99, 101, 129, 130,
    134, 151, 164, 170
Haskins, C. 165
Hay, D. 82
Haynes, R. 51, 54
Health Education Council 50
Henderson, J.A. 153
Henry, J.D. 17

Heseltine, M. 167, 171
Hetherington, P. 172, 174
Hillier Parker Research 13, 110
Hogg, S. 42
House of Commons 9
House of Commons Expenditure
    Committee 87
House of Lords Select Committee on
    Overseas Trade 160
Howe, G.M. 51, 54, 55, 145
Howells, J.R.L. 15
Hoyland, P. 176
Hudson, R. 102
Hughes, C. 17, 172

Independent, The 9
Issacharoff, R. 139

Jacques, M. 177
Jaskowiak, C. 126
Johnston, R.J. 39, 60, 62
Jones, D.R. 54, 146

Kay, J.A. 109
Keeble, D. 8, 9, 12, 20, 65, 67, 70, 105,
    109, 134, 136
Kelly, T. 9
Keogh, G. 109
Kieve, J.L. 89, 166, 167, 177
Killick, T. 67
Kitson, M. 12, 110, 136, 158

Labour Party 172, 183
Langridge, R. 9
Lee, C.H. 4
Lever, W.F. 109, 136
Lewis, J. 174
Liniecki, A. 22
Lock, D. 46, 82, 83, 105, 142, 144, 149,
    179
London Research Centre (LRC) 142, 143
Luttrell, W.F. 171, 180

MacInnes, J. 110
Maclennan, D. 65
Macmillan, H. 97
McCormick, B. 123
McCrone, G. 65
McGhie, C. 104
McQuaid, R. 13, 15, 129, 164
Malcolm, J.F. 109
Manners, G. 126
Manpower Services Commission 108
Marquand, J.M. 67, 83, 179
Marsh, P. 9

Martin, R.L. 2, 4, 5, 7, 9, 10, 12, 25, 68, 70, 76, 97, 98, 160
Mason, C.M. 18, 79
Massey, D. 82, 83, 99, 109, 136, 159, 161, 170, 180
Meegan, R.A. 109, 136
Merrett, S. 139, 170
Metcalf, D. 123
Mills, L. 85, 108, 109, 123, 136
Ministry of Labour 4
Minns, R. 139
Mintel Publications Ltd. 54, 57, 62, 147, 153, 154
Monk, S. 12, 110, 136, 158
Moore, B. 8, 12, 68, 70, 71, 72, 110, 136, 158, 163
Morris, A. 153
Muellbauer, J. 46, 47
Murie, A. 49, 62, 143
Murphy, A. 47

National Federation of Housing Associations 167
Nationwide Anglia Building Society 142
Naughtie, J. 26
Nellist, J. 166
New Society 106
Nicholson, G. 89
North of England Regional Consortium 83

O'Farell, P.N. 103
Oakley, R. 163
Organ, J.D. 15
Osmond, J. 2, 103, 120, 182
Owen, D.H. 12, 22, 41, 76, 139

Paddison, R. 153
Palmer, R. 105
Parr, J.B. 65
Parsons, D.W. 97
Pattie, C.J. 39, 60, 62
Peck, F.W. 68, 103, 110
Pond, C. 32
Prais, S.J. 109

Randall, J.N. 15
Regional Studies Association 17, 18, 70, 105
Reward Group 150
Rhodes, J. 8, 20, 46, 68, 70, 71, 72, 110, 136, 142, 158, 163
Richardson, R. 123
Ridley, N. 91
Riley, D. 10, 25, 26, 161

Roberts, J.L. 53
Robson, B. 79, 161
Roger Tym and Partners 89
Rogerson, R. 153
Rothwell, R. 163
Rowthorne, R.E. 25
Royal Town Planning Institute 181

Schofield, A. 8, 20, 37, 45, 98, 102, 126, 159, 160
Segal Quince and Partners 9
Shepley, C. 179, 181
Short, J. 139
Smallwood, C. 162
Smith, I. 174
Smith, M. 76
Spence, N.A. 26
Steer Davis and Gleave 22
Social Democrat Party 171
Southall, H. 2
Stott, M. 82, 106, 107, 177
Sunday Times 9, 153

Taylor, C.P. 50, 144, 145, 162
Taylor, J. 64, 65, 70, 71, 74, 76, 77, 161, 165
Thunhurst, C. 145
Tirbutt, S. 177
Town and Country Planning Association 1, 2, 4, 16, 17, 20, 28, 70, 71, 74, 102, 122, 142, 161, 162, 164, 169, 180
Townsend, A.R. 9, 12, 67, 68, 71, 76, 103, 104, 110
Townsend, P. 51
Travis, A. 76
Tyler, P. 8, 9, 20, 22, 25, 37, 46, 68, 70, 71, 77, 110, 136, 142, 158, 160

Vickerman, R.W. 20
Vincent, J. 17

Watts, D. 17
Webber, R. 129
Whitehead, M. 51, 146
Whittington, R.C. 20
Williams, A. 102
Williams, I. 103, 126
Wilsher, P. 102, 150, 153
Winter, P.D. 51, 54, 144, 145
Wintour, P. 79
Wood, P.A. 72, 74, 82, 94, 95, 105, 107, 111, 117, 136, 173, 179
Wray, I. 71, 83

Young, H. 57   Young, K. 85, 108, 109, 123, 136

SUBJECT INDEX

Aberdeen 12, 37, 103, 153; airport 20
Aberystwyth 147
accountancy firms 16, 137
Action for Cities (1988) 92, 94
activity rates 28, 65, 70
advertising firms 16
AEC 108
agriculture 28
air services 163
airports 163-4, 171
alcohol abuse 54, 165
Aldershot and Farnborough 56
Alliance 9, 60, 62
Ashford 164
Assisted Areas 8, 66-8, 70, 72, 74, 76-7,
    83, 91, 163, 165, 172
Association of District Councils
    (ADC) 171
Atlanta 169
atomic energy establishments 81
Attlee administration 97
atmospheric pollution 54
Aylesbury 56

Babergh 133
balance of payments 47, 82, 160
balance of trade 160, 172
banking 4, 99, 137; firms 16
Barlow Report (1940) 65, 181
Barnet 145
Barnsley 105
Basingstoke 56
Bathgate 56
Belgium 114
Berkshire 9, 17
Bethnal Green and Stepney 114
Beverley 151
Birkenhead 37; Birkenhead and
    Wallasey 56

Birmingham 17, 85, 91, 105, 158
Bishop Stortford 56
Blackpool 158
Board of Trade 65
Bolton 60, 158
Boothby, R. 97
Bosch 163, 174
Boston 169
Bournemouth 117
Bracknell 56
Bradford 92
branch firms/plants 5, 15-17, 71, 160,
    163
Breckland 133
Brent 123
Brighton 82, 117
Bristol 60, 92, 117, 146, 179
British Investment Bank 183
British Leyland 67
British Rail 81, 89, 183; Network
    SouthEast 183
Brittan, L. 77
Brixton 91
Brown, G. 26
Brussels 22; summit (1988) 168
Buckinghamshire 35, 114, 133
Budget (March 1988) 79, 161
Bury 60
business consultancy schemes 76
business expansion 160
Business Expansion Scheme (BES) 19, 79
business registrations 18, 33
business services 7, 16-17, 22, 46, 106,
    111

Caithness 51
Callaghan administration/government 70,
    98
Cambridge 9, 12, 15, 57

Cambridgeshire 35, 54, 133
Canada 165
Canary Wharf 111
capital gains tax 79
capital grants 72
capital restructuring 136
Caradon 133
carcinogenic hazards 54
Cardiff 15, 165; Cardiff Bay 89
Census of Employment (1984) 10
Census of population (1981) 149, 151
Central Electricity Generating Board 9
central-government localism 98
Channel tunnel 20, 22, 57, 144, 164, 183, 184
Chapeltown 92
Cheadle 150
chemical industry 71; chemicals 70
Chesham and Amersham 144
Cheshire 126; Cheshire Plain 153; North Cheshire 158
Chester 102
Chicago 169
Churchill administration 97
City Action Teams (CATs) 91-2, 94, 169
city grants 92, 94
Cleveland 17, 35, 36, 41, 133
Clitheroe 151
Clydeside 87
Coal 2, 9, 65, 87; coalmining industry 8-9
Coalition government vii
Coatbridge and Aidrie 56
Commission of the European Communities (CEC) 26, 168
community charge 166, 167, 172, 177
Community Development Projects (CDPs) 84
commuting 139, 154; fares 177
Comprehensive Community Programmes (CCPs) 84
Confederation of British Industry 22, 181; Employee Relocation Council 162
congestion 6, 65, 82, 109, 136, 139, 154, 164-5, 179-80
Congleton 151
Conservative Government/Party vii, 5-6, 10, 26, 46, 57, 60, 62, 63, 67, 74, 77, 87, 95, 97-9, 154, 158, 171, 172
Consett 39, 56
constrained location theory 134
consumer boom 126, 160
consumer expenditure/spending 7, 79, 126
Control of Offices and Industrial

Development Act (1965) 66
Cornwall 1, 30, 41, 101, 133, 145; North Cornwall 117, 133
corporation tax 166
cost of living 32
cotton textiles 2, 65
Council for the Preservation of Rural England 105
council house sales 49, 62, 79, 143
Council of Europe 107
county councils 172, 180
Coventry 92, 105
Crawley 15, 57
Croydon 145

Dagenham 145
Darlington 158
Deal 101, 147
decentralization 25, 72, 84, 100, 108-111, 134, 136, 165, 171, 178; jobs 17, 107; policy 71
defence equipment expenditure 80, 82, 99, 163
defence industries 134
deindustrialization vii, 4-5, 12-13, 28, 76, 83, 98, 102, 107, 109, 160, 172
Department of Employment ix
Department of the Environment 83, 85, 87, 91-2, 94, 144, 169, 171, 172; Secretary of State 142
Department of Health 77, 146, 165
Department of Social Security 77, 165
Department of Trade and Industry 46, 76-7, 83, 87, 92, 158, 171, 180-1
Department of Transport 171
dependency culture/economy 4, 95
depression vii, 4, 65, 68, 95, 97-8, 102, 120
Deprivation 101-2, 150, 159; urban 84-5, 91, 95
Deptford 92
Derby 163
Derbyshire 1-2
derelict land grants (DLGs) 91
Derelict Land Clearance Areas 91
deurbanization 133
Development Areas 25, 66-7, 74, 76, 163
devolution 171
Devon 1-2, 30, 133, 145
diet 54, 165
Disraeli, B. 97
Distribution of Industry Act (1945) 65, 97
Distribution of Industry Act (1950) 66
Distribution of Industry (Industry Finance) Act (1958) 66

district councils 171-2
Doncaster 92
Dorset 35, 36, 133
Dumfries 103; Dumfries and
  Galloway 103
Dundee 153
Durham 170; County Durham 15, 41,
  133, 163
Dyfed 9

earnings 29-30, 32, 99, 110, 149, 161;
  gross weekly earnings 30
East Anglia 1-2, 4, 12-13, 22, 25, 28, 29,
  35, 36, 38, 40, 41, 56, 64, 79-80, 103,
  105, 114, 117, 150, 179
East Midlands 1, 10, 12, 22, 34-5, 40-41,
  47, 49, 56, 62, 104, 139
East Sussex 36, 132
Economic Expansion Committees 179
Eden, A. 97
Edinburgh 102, 145, 153; airport 20
educational attainment 37
Edwards, N. 173
electronics 103
Emilia Romagna 180
employment vii, 6, 8-10, 12, 15, 17-18,
  20, 26, 28, 37, 55, 64-5, 67, 70-2, 80,
  84, 89, 92, 94-5, 97-8, 102-9, 111,
  114, 117, 123, 134, 136, 150, 159-60,
  162, 168-9; 177-8; banking 99;
  business 16; distribution 16;
  finance/financial services 16, 99; hotel
  and catering 16; insurance 99;
  manufacturing 4, 8, 10, 17, 45, 99,
  103, 105-9, 111, 117, 134, 177-8; part-
  time 107; public
  administration/sector 16, 160; service
  sector/services 4, 8, 10, 15-17, 106,
  110, 126, 137, 177-8
employment office area (EOA) 104
English House Condition Survey
  (1981) 54
English Industrial Estates Corporation 76
enterprise culture/economy vii, 4, 20, 76,
  95, 99
enterprise zones 89, 94, 98, 169
Essex 105, 114, 144
European Agricultural Guidance and
  Guarantee Fund 168
European Commission
see Commission of the European
  Communities
European Community (EC) 82, 112, 120,
  123, 125, 126, 168, 171, 183
European Regional Development Fund

(ERDF) 10, 26, 74, 168; grants 74
European Social Fund 168
Eurotunnel 164
exchange controls 4
Exeter 153
Expanded Towns 84, 136
Expanded Towns programmes 144

Falklands War 56
Falmouth 101; Falmouth and
  Camborne 117
Finance Act (1980) 89
finance 99; financial services 4, 12,
  16-17, 22, 106, 111, 178, 182
Firestone 108
fiscal measures/policies 6, 65, 163, 168
Folkestone 117
Fordist conditions 83
Forest Heath 133
France 179-80
Friends of the Earth 94
functional urban regions (FURs) 82
Fujitsu 163

Gatwick airport 12, 20, 163-4
General Elections (1979) 57, 67;
  (1983) 57; (1987) 9, 57, 59
gentrification 170
Glasgow 17, 145, 165; airport 20;
  City 53; Glasgow East Area Renewal
  (GEAR) 85-6; Greater Glasgow
  Health Board 145
Gloucester 9, 53
Golden Triangle 7, 22
Granby 92
Grampian 9, 103, 126
Great Yarmouth 101, 117
Greater London
  see London
Greater London Council viii, 82, 95, 107,
  176, 179
Greater London Enterprise Board 168
Greater London House Condition Survey
  (1985) 143
Greater Manchester
  see Manchester
Greater South East 105
Green Belt 82-3, 102, 105, 110, 144, 176
Greenwich 108
gross domestic product (GDP) 97; per
  capita 29, 46, 123, 125, 126, 173
Guildford 2, 39, 153
Gwent 133, 176

Hackney 1, 108, 143

Halifax 153
Hamburg 125
Hampshire 9; North East Hampshire 146
Handsworth 92
Handwerk associations 180
Hardman Report (1973) 67
Haringey 39, 139
Harrogate 1, 151, 153, 158
Hart 53
Hartlepools 56, 92, 153
Harwich 114
Hayes 108
head offices/headquarters of
    companies 7, 15, 17, 20, 71, 163
health 7, 51-5, 64, 144, 146, 149, 151,
    159, 165; authorities 173, 180;
    care 54, 137; services 177, 181
Health Education Council 51
Heath administration/government vii, 97
Heathrow airport 12, 20, 108, 163-4, 183
Hereford and Worcestershire 1
Hertfordshire 9, 144
Heseltine, M. 78, 182
Hexham 151
Highland Enterprise 184
Highlands 101; Highlands and Islands
    Development Board 184
high technology activity/industry 2, 4-5,
    9, 12-13, 15, 17-18, 20, 81, 101, 103,
    161, 163
High Wycombe 56
Highfields 92
Hitachi 174
Holderness 133
Holyhead 147
Home Counties 153, 172
home improvement 42
Home Office 84, 85
homelessness 94, 143, 170
Hoover 108, 174
Horsham 56
horticulture 28
hospital waiting lists 146
Hounslow 108
house-builders 142
House Builders' Federation 139
house-building 47, 48, 82, 84, 139, 142,
    165
house prices/values 6-7, 39-40, 42, 45-6,
    48, 61, 74, 126, 137, 139, 142, 151,
    161-2, 166-7; house price/incomes
    ratios 137; house price variations 98
household expenditure 31
housing 39, 41, 45, 46, 47, 53, 64, 84, 94,
    123, 137-9, 143-5, 149, 151, 159, 162,

    165-6, 169, 171, 176-9, 180;
    allowances 167; crisis 144; market 42,
    46; need 39, 143; rehabilitation 84-5;
    stock 41-2, 46, 48, 50, 84-5;
    subsidies vii, 94, 167, 169; unfit
    housing 143
Housing Acts: Housing Act (1969) 84,
    170; Housing Act (1974) 84, 170;
    Housing (Homeless Persons) Act
    (1977) 144; Housing Act (1980) 47,
    84, 170; Housing and Building Control
    Act (1984) 47; Housing and Town
    Planning Act (1986) 92; Housing Act
    (1988) 162
Howe, G. 98
Hull 182
Humberside 133
Huntingdon 133

Ile de France 177, 184
impannatori 180
improvement areas 84, 85
improvement grants 79, 170
incomes 4, 29, 31, 64, 84, 98, 159, 178;
    disposable incomes 102, 151; income
    distribution 7; per capita disposable
    income 29; personal incomes 30-1
income tax 79, 161
Index of Local Economic Performance 56
Industrial Development Act (1966) 66
industrial development certificates
    (IDCs) 66-8, 70, 72, 83, 85, 89, 136
Industrial Revolution 2, 12, 101, 145
Industrie und Handelskammern 180
Industry Act (1972) 66
Industrial Transference Scheme (1980) 65,
    97
information technology 4, 18
inheritance 45
inner area programmes (IAPs) 85
Inner Area Studies (IASs) 85
inner cities 6, 26, 67, 83-4, 87, 89, 91-2,
    94-5, 101-2, 134, 159, 168, 169, 177-8;
    inner city policy 85, 87, 94-5
inner city partnership programmes
    (ICPPs) 85
Inner Urban Areas Act (1978) 85, 91
insurance firms 16, 18
Intermediate Areas 25, 66-7, 74, 76
Inverclyde 54
Inverness 103
Island Areas 103
Isle of Wight 26, 36, 117, 133
Islington 145
Italy 114, 126, 180,

Japan 173, 174
job grants 72
John Lewis Partnership 126

Kendal 151
Kensington and Chelsea 150
Kent 20, 22, 101, 105, 114, 117, 151, 164
Keynesian policies vii, 5, 7, 67, 70, 98;
    incentives 65; inverted Keynesianism 5
King's Cross 22, 111
Kingston upon Thames 145
Knowsley 53
Knutsford 151

Labour Government/Party vii, 6, 9–10,
    26, 59, 61, 63, 95, 97–8, 100, 154, 158,
    183
Lambeth 85; West Lambeth 146
Lancashire North 158; Lancashire
    Enterprises Ltd. 169
land 65, 85, 104–5, 110, 142;
    acquisition 178; building land 142;
    costs 104, 136; derelict land 92, 174;
    industrial land 12–13, 158; supply 102;
    values 104, 111, 139
law firms 16
Leeds 15, 17, 92, 104, 158, 171; North
    East Leeds 158
Leicester 92
leverage 95; finance 89–90; ratios 92
Lewisham 92, 108
Liberal Party 57
Lincolnshire 133
Liverpool 1, 17, 37, 56, 85, 91, 94, 158,
    182
Llanelli 42
Local Employment Act (1960) 66
local enterprise boards (LEBs) 169–70,
    179, 181
Local Government Act (1972) 91, 169
Local Government Finance Act
    (1988) 166
Local Government, Planning and Land
    Act (1980) 89, 90
Local Government (Social Need) Act
    1969 84
local income tax 167
local labour market areas (LLMAs) 41,
    56–7, 139, 147, 153
local plans 82
Location of Offices Bureau (LOB) 66, 71
location quotients (LQs) 9, 15
Lochaber 51, 54
Lochalsh 54
London 2, 15–17, 19, 22, 41, 45, 56,

65–67, 71–2, 78, 82, 85, 91–2, 98, 102,
    105, 107–11, 120, 122–3, 126, 136–7,
    142–5, 147, 149–50, 154, 158–9, 164–5,
    167, 171, 176, 177, 183; City of
    London 18, 99; Greater London 12,
    20, 22, 25–6, 28, 30, 39, 40, 42, 48, 50,
    59, 66, 101, 108–11, 114, 117, 120–2,
    129–30, 134, 137–9, 142–5, 147, 150,
    166, 178; Inner London 82, 87, 102,
    108–9, 120, 122–3, 129–30, 133–4,
    143–4, 150, 158, 176, 177; London
    boroughs 120, 177; London
    docklands 20, 89, 170; London
    Docklands Development Corporation
    (LDDC) 89; London Housing
    Unit 144; London (Inner) South 120;
    London North East 120; London
    Regional Transport 81; London's third
    airport 99; Outer London 104, 120,
    129, 130
long boom vii, 4–5, 7–8, 67, 105, 109
Low Pay Unit 149
Lowestoft 117
Lowlands 103
Luton 114, 117
Luxemburg 113
Lyon 22
Lytham St. Annes 151

3M 174
Macclesfield 1, 151
Macleans 108
Macmillan, H. 97
macroeconomics 5; macroeconimic
    policies 6, 65, 98, 123, 160, 171
Maidenhead 15, 56
Maidstone 153
Manchester 17, 89, 91–2, 104, 153, 158,
    164; Greater Manchester 26, 133;
    (Ringway) airport 20, 163–4
manufacturing industries 4–6, 8–9, 12,
    15, 20, 26, 66–8, 70, 72, 99, 105,
    108–10, 117, 134, 136, 159–61, 167;
    base 160; exports 12; imports 12;
    labour 67; output 6, 68, 80, 98–9,
    103, 158; plants 67
    see employment
Matlock 151
mechanical engineering 2
Medway towns 101, 105, 153
Merseyside 17, 26, 35, 87, 89, 104, 133;
    Merseyside Development
    Corporation 89, 92; Merseyside
    Enterprise Board 169
metal industries 2, 70, 103;

manufacturing 71
Metrocentre 32, 126, 153
Meteorological Office 81
metropolitan counties vii, 25, 95
Mexborough 56
Mid-Glamorgan 36, 133, 163, 176
Middlesborough 51, 92
Midlands 1, 8, 20, 66, 105
migration 6, 35-6, 45, 129, 133, 139, 159,
    161, 177; in-migration 35, 123, 130,
    133; out-migration 8, 46, 70, 123, 130,
    136
Milton Keynes 9, 37, 56, 117, 147, 165
Ministry of Defence 80;
    establishments 80
Ministry of regional Development 180-1
Miskin 163
monetarism 5, 169; monetarist
    philosophy/policies vii, 77; monetary
    measures/policies 6, 65, 97
money supply 5
Monsanto 174
morbidity and mortality rates 55
mortality: acute myocardial
    infarction/heart disease 51, 53, 54,
    145; bronchitis 51; cancer 51, 54,
    144-5; melanoma 145; suicide 145
    see also standard mortality ratios
    (SMRs)
Morpeth 151
mortgage interest rates 126, 137, 138;
    mortgate interest relief 79, 82, 167
Motherwell 54
motorways 74, 136, 161; M1 164;
    M2 164; M3 corridor 12, 15, 20, 108,
    134; M4 corridor 9, 15, 18, 20, 108,
    134; M6 164; M11 corridor 20, 99,
    144; M25 82, 99, 144, 164; M40 144;
    M62 182
Moss Side 91-2
multinational
    capital/companies/corporations 7, 8,
    17, 99, 136, 170, 171
Munich 22

National administration/government vii,
    97
National Health Service 146, 177
National Physical Laboratory 81
National Regional Development
    Council 181
Netherlands 114
New Earnings Survey 149
New Towns 2, 105, 136; New Town
    programmes 144

New Towns Act (1946) 84; (1959) 84
Newbury 56
Newcastle 17, 91, 158; airport 20; Eldon
    Square Centre 126
Newham 108, 145, 150, 170
Newton Aycliffe 15, 163
Nissan 73, 153, 163
North/Northern Region 2, 6, 13, 17, 22,
    25-6, 32, 35-6, 51, 54, 56, 62, 68, 74,
    104, 113-14, 120-21, 134, 137, 139,
    150-51, 153, 164, 166, 181
north-north divide 32
North of England Development
    Council 181
north: Inner Urban North 129-30, 134;
    Outer Urban North 130, 134; Rural
    and Resort North 130, 134; Suburban
    North 130, 134
north-south disparity/divide/division vii,
    viii, 1, 4, 9, 12, 22, 25, 28-9, 36, 40,
    45-6, 48, 50, 56, 59, 61, 64, 79, 98-9,
    101-2, 105, 112, 114, 126, 137, 146-7,
    151, 153-4, 158-9, 164, 166, 171, 173,
    180
North West 1, 6, 10, 15, 22, 25-6, 34-5,
    56, 61, 68, 80, 114, 151, 164, 166, 183
Northamptonshire 1
Northern Development Company
    (NDC) 183
Northern Ireland viii, 6, 22, 26, 37,
    39-40, 45, 56, 64, 113, 125, 137
Northern Ireland Office 171
Norfolk 133
North Peckham 92
North Yorkshire 133
Northern Lights 151
Norwith 117
Notting Hill 92
Nottingham 92

offices 66-7; government offices 165
office development permits (ODPs) 66,
    71-2, 85, 136
oil 67; industry 28, 71, 103; price hike 97
Organization for Economic Cooperation
    and Development 98
Orkney 9
Oxford 9, 54

Panasonic 174
Paris 22, 177
Park Royal 108
partnership areas 85
pension funds 78, 167, 169
Penzance 101

Peterborough 117
Peterlee 15
Pickfords 126
Plaid Cymru 57
political polarization 59
population 25, 34-6, 55, 79, 91, 108, 117, 130, 133-4, 136, 142-3, 146, 150, 158, 168, 181; working population 67, 74, 97, 114, 177
Portsmouth 105, 117
Portugal 125
Potteries 1
Powys 9, 133
Prato 180
premature equity release 42
Preston 92
production cost explanation 136
programme areas 85
prosperity 56; divide 56
public administration 16, 107, 165
public expenditure/spending 8, 67, 76-7, 82, 95, 99, 160-61, 163, 165, 168, 170, 174, 177, 178
Public Health Act (1872) 54; (1875) 54
public transport subsidies 82
Pyrene 108

quality of life 145, 151, 153-4, 159, 165, 177; index 150

Radnor 133
rail services 163-4
rates 166, 177; rate capping viii; rate subsidies 65; rate support grant viii, 94, 169
recession 5, 8, 13, 26, 46, 56-7, 68, 70, 72, 76, 95, 98-9, 110, 134, 160, 163
redundancy 68
regional audit 165
regional councils 172, 176, 180, 182-3
regional development agencies 171-2, 176, 179, 182
regional development grants (RDGs) 66-7, 70-2, 74, 76-7, 163
Regional Development Grant (Termination) Act (1988) 77
regional economic planning boards 97; councils 97
RER network 177, 179
regional employment premiums (REPs) 66, 71
regional enterprise grants 76
Regional Impact Analysis 181
regional investment banks 172
regional selective assistance 70, 76

rent subsidies 65
rents 137; housing 143; industrial 12-13, 110, 134, 165; office 66, 134, 137; shop 126
research and development 7, 9, 15, 46, 71
research establishments 12, 15
retail: centres 153; prices 139, 142; trade 123, 150; retailing 137, 144
Ridley, N. 91
Ripon 151
road pricing 177
Road Research Laboratory 81
Rochdale 92
Rochford 54
Rotherham 105, 126, 170
Rover 67
Rutherford Laboratory 81
Ryedale 133

St. Ives 117
St. Paul's 91, 92
San Francisco 169
Sandwell 89
Schedule A tax
Scotland viii, 2, 6, 13, 15, 17, 25-6, 28, 30, 40, 51, 56, 61, 71, 74, 81, 84-5, 102-4, 113-14, 120, 137, 150, 180; Scottish assembly 183; Scottish Development Agency (SDA) 85, 103, 179, 181, 183; Scottish Enterprise 184; Scottish Nationalist Party 59; Scottish Office 171, 180
SEEDS (the South East Economic Development Strategic Association) 82
Selby 133
SERPLAN (South Eastern Regional Planning Conference) 46, 142, 144, 176
service activity/industries 4-6, 9, 28, 110, 134, 165, 167
see employment
Severn 1, 2; Severn-Wash line 51, 54, 55, 89, 102
Sheffield 92, 104, 145, 158, 170; Hallam 158
shipbuilding/ships 2, 9, 65, 87
Silicon Glen 15, 101
Single European Act (1986) 168
single uniform business rate 166
site value rate 167
Skye 54
Slough 153
smoking 54, 165
Social Democratic Party 60, 182
Solihull 105

Somerset 133
Sony 174
South East 1-2, 5-10, 12-13, 15-20, 22,
    25-6, 28, 30-2, 35-6, 39-40, 45-6, 50,
    53, 55-6, 61, 64-7, 70-2, 79-83, 92,
    99-107, 109-10, 114, 117, 121, 125,
    129-30, 134, 136-9, 142, 144-7, 150,
    153-4, 159, 161, 163-7, 171-2, 172,
    176-9; Outer Metropolitan Area 22,
    40, 99, 129-30, 134, 143; Outer South
    East 41, 129-30, 134, 179; South East
    Regional Economic Planning
    Council 82, 176
south-south divide 28
South Shields 56
South West 1-2, 4, 12-13, 25, 35, 40, 50,
    57, 61, 80, 92, 103, 105, 117, 150, 179
South Wight 139
South Yorkshire 9, 17, 41, 133
Southall 108
Southampton 117
Southend 114
Southwark 1, 145
Spain 114
Special Areas legislation 97; Special Areas
    (Development and Improvement) Act
    (1934) 65; Special Areas
    Reconstruction (Amendment) Act
    (1936) 65; Special Areas (Amendment)
    Act (1937) 65
Special Development Areas 66, 163
Stafford 103
Staffordshire 1
standard of living 64
standard mortality ratios (SMRs) 51, 54,
    145
    see also mortality
Stansted airport 20, 144, 164
steel industry 9, 65, 71, 101
Strategic Plan for the South East 82
stockbroking firms 16
Stevenage 15
Structure Funds 168
structure plans 82, 144, 172
Suffolk 133
Sunderland 1, 37, 56, 72, 163
supplementary benefits 78
Sur 113
Surrey 9, 42; West Surrey 146
Sussex 117
Sutton 145
Swansea 153

task forces 91-2, 94, 169
tax privileges and exemptions 79

Tebbit, N. 77
Tees Bay Retail Park 153
Teesside 71, 89
Telford 165
tertiarization 4-5
textile industry/textiles 9, 87, 158
Thames estuary 37; Thames Valley 9;
    Thamesside 114, 117
Thanet 114, 151
Thatcher
    administrations/governments vii, viii,
    4-6, 12, 56, 62, 67, 72, 76, 78, 83,
    91-2, 94, 98-100, 159, 169, 171;
    Thatcherism 9-10
Thirsk 151
Torbay 117
Torquay 101
tourist industry 28
Tower Hamlets 92, 108, 170
Town and Country Planning Act
    (1947) 65-6, 97
Toxteth 91-2
Toyota 163
trading estates 66
Trafford Park 89
Transpennine 182
travel to work areas (TTWAs) 103-4, 151
Tweeddale 51, 54
Tyne and Wear 133
Tyneside 87, 89

unemployment vii, ix, 2, 4, 6-9, 22, 25-6,
    28, 29, 37, 39, 43, 46, 56, 67-8, 70, 77,
    84, 89, 94-5, 97, 99, 102-3, 105, 108,
    112, 114, 117, 120-23, 142, 149-51,
    158-63, 165, 170, 174, 177, 178
Unemployment Unit ix
urban development corporations
    (UDCs) 89, 92, 94, 98, 169
urban development grants (UDGs) 91-2
urban regeneration grants (URGs) 92, 94
urban programme (UP) 84-5, 91-2, 94,
    98
urban renewal 84, 163, 170
USA 165, 169, 172, 174

vacancies 34; vacancy rates 28
venture capital 5, 18

wages 46, 107, 136; national wage
    agreements 46, 98; national wage
    bargaining 99, 162; regional wage
    determination/settlements 98, 102;
    wage levels 63, 126
Wales viii, 2, 6, 13, 15, 17, 26, 40-1, 56,

62, 68, 74, 80-1, 85, 104, 114, 120, 139, 147, 172-3, 174-6, 182; Welsh Assembly 183; Welsh Development Agency (WDA) 174, 175, 177, 182, 181; Welsh Marches 158; Welsh Office 171, 180
Walker, P. 85, 172-3, 174
Warwickshire 1
Wash 1-2
Washington 153
water 172; industry 181, management 181; undertakings 183
west-east divide 151
West Germany 114, 126, 169, 174, 179-80
West Glamorgan 35, 36, 133, 174
West Midlands 1, 22, 26, 35, 39, 61, 65, 67, 70, 74, 89, 103-4, 164, 166; West Midlands Enterprises Board 169; West Midlands Industrial Development Association 103; West Midlands metropolitan county 133
West Sussex 133
Wimborne 133
Winchester 56

Whitby 51
White Papers: White Paper on Employment Policy (1944) 6, 65; Policy for the Inner Cities (1977) 85; Regional Industrial Development (1983) 72, 74; DTI - the Department for Enterprise (1988) 76; The Future of Development plans (1989) 172
Wigan 165
Wilson administrations/governments vii, 97-8, 174
Winchester 2, 9
Wolverhampton 92

York 103, 153, 170; Vale of York 158
Yorkshire 154; East Riding; Yorkshire and Humberside 1-2, 13, 26, 34-5, 56, 80, 99, 104-5, 137, 139, 151, 164, 166, 182; East Yorkshire 133; North Yorkshire 133; West Yorkshire Enterprise Board 169

Z score 147